Construction Contract Administration

AWWA MANUAL M47

First Edition

FOUNDED 1881

American Water Works Association

MANUAL OF WATER SUPPLY PRACTICES — M47, First Edition

Constuction Contract Administration

Project Manager and Technical Editor: Kathleen A. Faller
Technical Writer: Todd A. Shimoda
Copy Editor: Phillip Murray
Production Editor: Alan Livingston

Construction contract administration, M47. —1st ed.
xviii, 149 p. 17×25 cm. — (Manual of water supply practices ; 47)
Includes bibliographical references and index.
ISBN 0-89867-871-4
1. Water treatment plant construction contracts—United States.
I. American Water Works Association. II. Series: AWWA manual ; 47. KF865.C636 1996
628.1′62′0687—dc20

Printed in the United States of America

American Water Works Association
6666 West Quincy Avenue
Denver, CO 80235

ISBN 0-89867-871-4

Printed on recycled paper

Contents

List of Figures, vii

List of Tables, ix

Foreward, xi

Acknowledgments, xiii

Introduction, xv

Principal Terms, xvii
Publicly Owned Versus Privately Owned Utilities, xvii
Chapter Preview, xvii

Section I Contracting Issues **1**

Chapter 1 Traditional Construction Projects **3**

Project Phases, 4
Participant Responsibilities, 6
Consultant Selection, 10
Traditional Project Advantages and Disadvantages, 11
Reference, 12

Chapter 2 Construction Documents Review and Application **13**

Construction Documents and Contracts, 13
Engineers Joint Contract Documents Committee, 14
References, 24

Chapter 3 Warranties **25**

Participant Responsibilities, 25
General Content, 26
Conditions and Requirements, 30
Claims, 35

Chapter 4 Prequalifying and Selecting Contractors **37**

Prequalification, 37
Competitive Bidding Process, 42
References, 44

Chapter 5 Performance Clauses **45**

Types of Performance Clauses, 45
Other Performance Incentives and Issues, 47
Examples, 48

Section II Nontraditional Construction Contract Administration . . . **51**

Chapter 6 Owner Construction Management **53**

Characteristics and Activities, 53
Staff Requirements and Responsibilities, 55
Advantages, 58
Disadvantages, 60

Chapter 7 Construction Management Firms **61**

Characteristics, 61
Advantages and Disadvantages, 63
Responsibilities and Procedures, 64

Chapter 8 Multiple Prime Contracts **69**

Multiple Prime Contracts Project Management, 69
Characteristics of MPC Projects, 72
General Qualifications and Procedures, 76
Reference, 77

Chapter 9 Design–Build and Turnkey Construction **79**

Design–build, 79
Turnkey Construction, 83
Standard Documents, 84
Reference, 85

Chapter 10 Guaranteed Maximum Price Construction Management . . **87**

Characteristics, 87
Objectives of GMPCM, 89
Participant Responsibilities, 91
Reference, 96

Section III General Issues **97**

Chapter 11 Partnering **99**

Case Studies, 101
Reference, 104

Chapter 12 Communication **105**

Basics of Good Communication, 105
Project Communication, 108

Chapter 13 Disputes **111**

Background, 111
Dispute Resolution, 115
Alternative Dispute Resolution, 120

Chapter 14 Payment Procedures **123**

Payment Methods, 124
Payment Schedule, 125

Important Elements, 126
Payment Problems, 128
Project Completion, 129

Chapter 15 Private-Sector Financing **131**

Privatization, 131
Strategies to Attract Private-Sector Financing, 133
Public-Sector Financing Process, 133
Situations Suited to Private-Sector Financing, 135
Leasing, 138

Bibliography, 141

Index, 143

List of AWWA Manuals, 149

Figures

I-1 Large waterline installation, xv

I-2 Construction of a 12-mgd (0.5258-m^3/s) nanofiltration water treatment plant in Fort Myers, Fla., xvi

I-3 Newly installed polymer feed pump, xvi

1-1 Traditional project participants, 3

1-2 Installed air-gap type of backflow preventer with valves, 4

1-3 Conventional water treatment plant (4 mgd [0.1753 m^3/s]) in Quartz Hill, Calif., 5

1-4 Interior of a well-supplied project engineer's field office, 8

2-1 Typical contents of bidding documents, 14

2-2 Typical contents of contract documents, 15

2-3 Severe washout damage at a treatment plant, 22

2-4 Stabilization of creek banks with rock-filled gabions, 23

2-5 Equipment used in creek-bank repair, 24

3-1 Installation at 12-mgd (0.5258-m^3/s) nanofiltration water treatment plant in Collier County, Fla., 27

3-2 Installation of tee for a 60-in. (1,500-mm) water main, 28

3-3 Chlorine decontamination equipment, 29

3-4 Pipe stored on a busy street, 31

3-5 Handling a 60-in. (1,500-mm) waterline, 32

3-6 Placing an 80-in. (2,000-mm) diameter steel pipeline in suburban Washington, D.C., 33

3-7 Sixty-in. (1,500-mm) waterline connections, 34

3-8 Sixty-in. (1,500-mm) waterline backfill, 35

3-9 Operator cleaning the socks of a bag-type dust collector installed to prevent loss of lime dust to the atmosphere, 36

4-1 Tunnel for a 96-in. (2,400-mm) diameter steel pipeline in suburban Washington, D.C., 38

4-2 Seventy-mi (112.6-km) pipeline consisting of 54- and 30-in. (1,350- and 750-mm) diameter pipe, Yucca Valley, Calif., 39

4-3 Full-circle repair clamp on older waterline, 40

6-1 Steelworkers on 6-mil gal (22,710-m^3) reinforced concrete tank in San Diego County, Calif., 54

6-2 Stages in the construction of an elevated steel tank needed to maintain the necessary head for filter backwash water, 58

6-3 Expansion of conventional water treatment plant from 3 to 6 mgd (0.1314 to 0.2629 m^3/s) in Auburn, Calif., 59

7-1 Pump station (100 mgd [4,381 m^3/s]) in Orange County, Calif., 62

7-2 Fort Myers, Fla., 12-mgd (0.5258-m^3/s) nanofiltration water treatment plant, 62

7-3A Installation of 54-in. (1,350-mm) discharge line in chlorine contact structure, 65

7-3B Installation of lines in chlorine contact structure, 65

7-3C New discharge pipes in chlorine contact structure, 66

8-1 Typical flow of information for multiple prime contractors, 70

8-2 Typical construction management flow of information, 71

8-3 Foundation work on dechlorination building, 73

8-4 Finish work on expansion of conventional water treatment plant from 6 to 12 mgd (0.2629 to 0.5258 m^3/s), 74

9-1 Elevated steel reservoir (0.75 mil gal [2,839 m^3]), City of Richardson, Texas, 80

9-2 Two 16-mil gal (60,570-m^3) buried prestressed concrete reservoirs in Orange County, Calif., 81

9-3 Roller-compacted concrete dam (139-ft [42.4-m] high) in Ruidoso, N.M., 82

10-1 Fort Myers, Fla., 12-mgd (0.5258-m^3/s) nanofiltration water treatment plant, 89

10-2 Pump station (18.25 mgd [0.7996 m^3/s]) in suburban Washington, D.C., 90

10-3 Foundation work completed (note silt fence in foreground), 91

10-4 Bank of feed pumps, 95

11-1 Partnering agreement, 100

11-2 Typical agenda for a partnering workshop, 103

12-1 Communication model, 105

12-2 Construction signage displaying a concise message, 106

12-3 Contrast in communication., 107

13-1 Typical contractual arrangement 30 years ago, 112

13-2 Typical general contractor format today, 113

13-3 Typical format for modern large projects, 114

13-4 Typical modern format with multiple prime contractors, 115

14-1 Sample payment procedure work flow, 127

14-2 Sample contractor payment work flow, 128

Tables

6-1 Example of estimating project staff requirements, 55
15-1 20-year maturities—borrowing cost, 132

Foreword

This manual introduces the terms and procedures used in water utility construction project administration in the United States. The intended audience is anyone involved in such projects, including engineers, contractors, suppliers, administrators, and decision makers. The main emphasis of the manual, however, is presented from the water utility's, or owner's, point of view.

Because construction projects vary so much from project to project and country to country, and state and local regulations vary nearly as much, this manual cannot address every specific issue. And because of the variance, the manual cannot be considered a standard of practice. It should be used, rather, as a resource in planning, designing, and managing construction projects.

Contract documents have important legal consequences, and consultation with a qualified attorney is encouraged before entering into a contract for the performance of construction, engineering services, construction management, and other services.

This is the first edition of the Construction Contract Administration Manual. The American Water Works Association (AWWA) and the Construction Administration Committee would appreciate any comments on the manual. Contact the AWWA Standards and Product Development Department at (303) 794-7711, 6666 W. Quincy Ave., Denver, CO, 80235.

Acknowledgments

The American Water Works Construction Administration Committee (593) of the Engineering and Construction Division developed this manual. Special appreciation is expressed to Richard S. Gabrielse, former committee chair, for initiating and following this project. Authors and reviewers are as follows:

D.W. Rommelmann, Chair, Boyle Engineering Corporation, Denver, Colo.
A.V. Almeida, Dallas Water, Dallas, Texas
R.D. Ambrose, Black and Veatch, Gaithersburg, Md.
W.F. Anton, Seattle Water Dept., Seattle, Wash.
H.H. Benjes Jr., HDR Inc., Dallas, Texas
W.M. Breichner, Mainlining Service Inc., Hagerstown, Md.
C.H. Carpenter, Provo City Dept. of Water Resources, Provo, Utah
J. Chinnock, URS Consultants, Denver, Colo.
W.J. Dixon, Dixon Engineering Inc., Lake Odessa, Mich.
P. Dobrolski, City of Houston, Houston, Texas
W.T. Flynn, City of Austin, Austin, Texas
T.A. Goebel, Past Chair, Hazen & Sawyer, P.C., Raleigh, N.C.
R.D. Giardina, Rick Giardina and Associates Inc., Englewood, Colo.
K.W. Henderson, Malcolm Pirnie Inc., White Plains, N.Y.
R.D. Hugus, Philadelphia Suburban Water, Bryn Mawr, Pa.
C.J. Jeckell, City of Virginia Beach, Virginia Beach, Va.
J.C. Jenkins, City of Adrian, Adrian, Mich.
J.E. Layne, Lakewood, Colo.
R.F. Miller, Indianapolis Water Co., Indianapolis, Ind.
J.P. Naismith, San Patricio Municipal Water District, Ingleside, Texas
R.H. Peterson, Municipal Research Services Center, Kirkland, Wash.
W.M. Piatt III, W.M. Piatt and Company, Durham, N.C.
G. Pippin, Tualatin Valley Water District, Beaverton, Ore.
G.A. Raftelis, Raftelis Environmental Consulting Group, Charlotte, N.C.
J.C. Reichenberger, Loyola Marymount University, Los Angeles, Calif.
R.M. Robler, City of Colorado Springs, Colorado Springs, Colo.
S.P. Schmitt, American Water Works Service Co., Voorhees, N.J.
D.G. Smith, Black & Veatch Engineers, Tampa, Fla.
W.R. Snapp, Wilbur Smith Associates, Columbia, S.C.
J. Stevens, Waterloo Water Works, Waterloo, Iowa
T.W. Stewart, Dallas Water Utilities, Dallas, Texas
N. Trainor, Camp Dresser & McKee Inc., Ft. Lauderdale, Fla.
T.W. Vandeventer, Charlotte Mecklenburg Utility Department, Charlotte, N.C.
R.K. Weir, Denver Water, Denver, Colo.

Photographs used in the manual were provided by the City of Austin, Texas, Water Utility (courtesy of William Flynn and Mohan Gadhia), Boyle Engineering Corporation (courtesy of David Rommelmann and Vaughn Shelton), the City of Fort Collins, Colo., Water Utility (courtesy of Terry Robinson), and L.J.C. Shimoda.

Introduction

Construction projects must be managed to be successful. Such project management is called contract administration because the agreements between the project participants are set in legal contracts. This manual presents guidelines for effective contract administration. These guidelines are based on accepted practices used in water utility construction and similar types of projects. For example, Figures I-1, I-2, and I-3 illustrate typical water utility construction projects.

Figure I-1 Large waterline installation

This introductory chapter discusses the principal terms used in the manual, explains the major differences in contract administration for publicly owned versus privately owned utilities, and previews the chapters of this manual.

Figure I-2 Construction of a 12-mgd (0.5258-m^3/s) nanofiltration water treatment plant in Fort Myers, Fla.

Figure I-3 Newly installed polymer feed pump

PRINCIPAL TERMS

The terminology used in contracts and construction varies from locale to locale. The terms used in this manual are not intended to be the only correct terms that can be used. For example, a project manager is sometimes called a resident manager, a project engineer, a construction manager, or a resident engineer. Depending on the project and locale, there may be differences in the meanings of the terms. The construction documents of a project should define the terms to be used on a specific project.

The term *owner* refers to the entity, usually a utility, that arranges the financing for the project, hires the other participants, and operates and maintains the project after it is completed.

The term *consultant* refers to a professional or a firm hired by the owner to perform planning, design, or construction observation services. In this manual, the consultant is also referred to as the *engineer* or *design engineer* (in reference to engineering-related services). Of course, a consultant may also perform services other than engineering, such as architectural design.

The term *contractor* refers to the firm or consortium that constructs the project. Other groups, such as the American Society of Civil Engineers, and the American Water Works Association (in its standards), refer to the *contractor* as the *constructor*. This manual uses the more common term, however. *Contractor* usually implies a general or prime contractor. Firms hired by the general contractor are *subcontractors*.

As previously discussed, *contract administration* is the term used in this manual to refer to the owner's project management efforts. Contract administration occurs over the complete project from planning through postconstruction. The term *construction management* is more specific to the construction phase of the project.

Other terms are defined as needed in the following chapters.

PUBLICLY OWNED VERSUS PRIVATELY OWNED UTILITIES

The two basic types of project owners are publicly owned utilities and privately owned utilities. There can be differences between the two in construction projects; for example, a private owner may not be bound to competitive bidding as strictly as a public entity. For the purposes of discussion in this manual, the owner is assumed to be either publicly or privately owned unless strong differences exist. In that case the differences will be discussed.

CHAPTER PREVIEW

The manual is presented in three main sections intended to divide the topic of contract administration into distinct areas of discussion. However, some overlap is occasionally necessary to provide a complete discussion.

Section I introduces the common contracting issues of construction projects. Chapter 1 discusses the traditional (owner–consultant–contractor) construction project, including the project phases and participant responsibilities. Chapter 2 introduces construction documents, which include bidding requirements, contracts, drawings, and specifications; the chapter also reviews standard contract documents developed by the Engineers Joint Contract Documents Committee. Chapter 3 clarifies project warranties, which are guarantees of the contractor's and suppliers' materials and work; warranties are specified in the contract documents. Chapter 4 discusses the steps in selecting contractors and the reasons for prequalifying contractors. Chapter 5 ends the first section with a discussion of performance clauses; these contractual clauses give the contractor incentives to complete the project on time.

Increasingly, drinking water supply projects are constructed using nontraditional participant arrangements. Section II introduces the commonly used nontraditional approaches to construction projects and contract administration. These approaches range from contract administration performed by the project owner (chapter 6) to turnkey projects (chapter 9) in which a private firm completes all the phases of the project. Other chapters in Section II discuss hiring a construction management firm to provide contract administration (chapter 7), projects with multiple prime contractors (chapter 8), and projects with a guaranteed maximum price (chapter 10).

Section III discusses general issues of contract administration including partnering (chapter 11), communication (chapter 12), contract disputes (chapter 13), payment procedures (chapter 14), and private sector financing (chapter 15). These general issues apply to all types of projects and contract administration.

At the end of this manual, the bibliography lists resources that can be consulted for more detail about the topics introduced in these chapters.

Section I

Contracting Issues

Traditional Construction Projects

Construction Documents Review and Application

Warranties

Prequalifying and Selecting Contractors

Performance Clauses

Chapter 1

Traditional Construction Projects

This chapter defines the traditional water utility construction project, project phases, participants and their responsibilities, selection of participants, forms of payment, and advantages and disadvantages of the traditional project.

In earlier times, more often than not, water utility staff would plan, design, and construct the utility's projects themselves. As projects became more complex, water utilities contracted with outside experts, particularly to design or construct the project. The utility continued to provide the planning, design review and approval, financing, bidding, and contract administration. More recently, the consultant may also perform the planning, bidding, and construction contract administration.

Traditional construction project organization can be illustrated by a triangle (Figure 1-1); the utility (owner) hires the design engineer or architect (consultant) and the company who builds the project (contractor). The consultant and the contractor

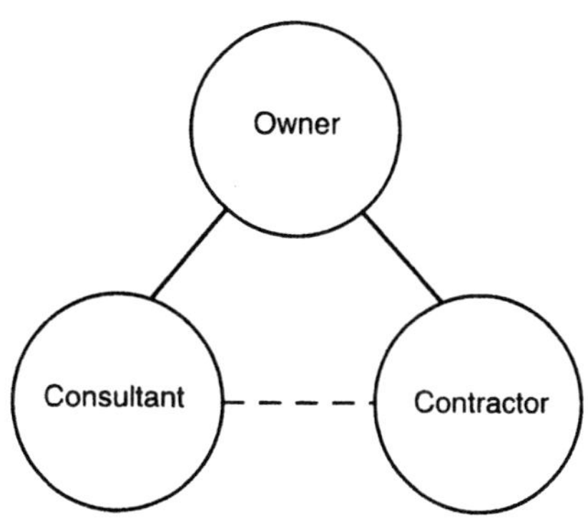

NOTE: Solid line indicates formal contract ties; dashed line means no formal ties.

Figure 1-1 Traditional project participants

are connected with a dashed line, symbolizing that those two participants do not have contractual ties but may interact during construction if the consultant performs the contract administration.

PROJECT PHASES

The following paragraphs introduce the main phases and activities of the traditional water utility construction project. The responsibilities of the participants during each phase are discussed in following sections. Of course, water utility projects vary widely, from installing a valve (Figure 1-2) to constructing a treatment plant (Figure 1-3), and the phases also vary according to the complexity of the project.

Planning

A preliminary study defines what the project needs to accomplish, determines possible alternatives to meet the need, estimates costs, and determines financial capability. Based on the preliminary study, a final report more accurately defines the project scope, suggests a preferred alternative, and presents a refined budget. The owner uses the preliminary study and final report to decide whether to proceed with the project. The final report also serves as the basis of the design.

Design

The major products of the design phase include construction drawings, specifications (technical guidelines), and the bidding documents (legal guidelines). Other design-phase products typically include a final construction cost estimate and agency approvals of the design.

Based on the construction cost estimate and financing, the type of construction contract is usually determined in this phase. The main types of contracts in traditional projects are fixed price (unit bid or lump sum) and cost-plus.

Figure 1-2 Installed air-gap type of backflow preventer with valves

Figure 1-3 Conventional water treatment plant (4 mgd [0.1753 m^3/s]) in Quartz Hill, Calif.

In a fixed-price, unit-bid contract, quantities for each element of the work are defined within the bid section of contract documents. During the bidding phase, bidders fill in the unit prices they will charge for each line item (for example, 100 lin ft of 6-in. pipe at $5 per foot). Some projects do not easily lend themselves to a measurable quantity definition. In those cases, areas or limits of work are defined and the contractor is requested to submit a lump sum bid. During construction, payment is made based on an agreed on percent completion of those elements appearing within an approved monthly cost estimate breakdown, also referred to as the schedule of values.

In a cost-plus contract, payment is made for the actual cost of the work based on labor and equipment rates, material invoices, and profit and overhead markup percentages. The effort is tracked during performance of the work. More prevalently, the cost-plus approach is used as part of a change-order process or on a project for which the limits of effort required cannot be reasonably defined and the contracting parties cannot agree on a lump sum value.

Bidding

The bidding phase essentially involves the selection of the contractor. Chapter 4 discusses this topic at length. Generally, the availability of bid documents (plans and specifications) to prospective bidders is advertised in local papers as well as within known construction publications, such as the *Dodge Reports Bulletin*. A prebid conference may be held at which time questions from prospective bidders are answered. Bids are received and evaluated for compliance with funding capability, conformance with procurement requirements, and qualifications of the bidders. This phase also entails awarding a contract and entering into an agreement.

Construction

During the construction phase, the project is built according to the drawings, specifications, and contract provisions. Of course, projects rarely go completely according to the plans because of unforeseen conditions, and changes may be required. All three participants—owner, consultant, and contractor—are usually involved in changes.

During construction, there are many points of communication, written and spoken, and much documentation is prepared. The types of documentation used include shop drawings, requests for information, responses to requests for information, payment requests, change orders, inspections, and material and operational test results. The components of the project are observed and either accepted or rejected. All components must be working correctly (as defined by the specifications) before the project can be considered finished.

Postconstruction

In the postconstruction phase, utility staff may need instruction in the operations and maintenance (O&M) of the project components. A warranty period (refer to chapter 3) is usually in effect.

PARTICIPANT RESPONSIBILITIES

The three project participants have relatively distinct responsibilities during each phase of the traditional construction project. Again, projects vary in complexity and the responsibilities vary accordingly. Each party's responsibilities should be spelled out in the contract documents.

Owner

The owner has responsibilities in all phases of the project, the most important of which are defining the need for the project and selecting the consultant and contractor. The contractor is usually selected through the bidding process as described in this chapter. Guidelines for selecting the consultant are presented in the "Consultant Selection" section of this chapter. The owner is responsible for clearly defining and communicating the expected roles of all the participants.

Preliminary. Based on the utility's needs, the owner defines the scope of the project. The owner, sometimes together with the consultant, performs the preliminary study and prepares the final report. The owner acquires project approval and funding from the necessary authority, such as a water utility governing board.

Design. The owner enters into an agreement with the consultant to generate drawings, specifications, and bidding documents. During the design phase, owner and consultant meet to review progress and resolve any questions that arise. The owner must make sure the project stays within its scope and budget and that the design meets expectations. The owner should provide necessary design approvals.

Bidding. The owner may submit the bid request to a list of prequalified contractors, advertise for bids, or both. Often, the owner is bound by certain legal bidding requirements. Local newspapers, as well as construction bulletins such as

Dodge Reports Bulletins, can be used to advertise. The length of the advertisement period may be based on local, state, or federal regulations; project complexity may also determine the length.

The owner, often together with the consultant, performs the bid opening, where those bidding are present and their bids are opened and read. The owner, again sometimes with the consultant, evaluates the bids; if the low bidder meets the criteria established in the bidding documents, the low bidder is selected to construct the project.

Construction. The owner issues a notice of award to the approved low bidder and sends the construction contract to the contractor to be signed. The owner also signs the agreement and issues a notice to proceed, which defines the start date and completion period. During construction, the owner documents progress and inspections, attends established progress meetings, approves appropriate changes to the project (working with the design consultant if the changes involve the design), and issues payments to the contractor. When the project is completed and accepted, the owner issues a project completion statement, the date of which establishes the beginning of the warranty period. The contract documents must define the terms of acceptance. Figure 1-4 shows the on-site office of a project engineer.

Postconstruction. During the latter stages of the project, as components become ready for their intended use, the owner's staff may need training in the operation of the components. The owner at this point becomes responsible for O&M of the project.

Consultant

In the traditional construction project, the consultant's main responsibility is to prepare the project plans and specifications. In recent years (particularly with more complex projects), the consultant's role has come to include tasks such as

- performing preliminary studies
- generating final reports
- performing financial studies
- conducting the bidding
- providing construction contract administration
- assisting with project startup
- assisting with O&M

Preliminary. The owner may hire the consultant to perform or assist with the preliminary study and final report. The consultant often estimates the construction costs and may present design and financing options. The consultant may assist the owner with obtaining regulatory agency approval as well as approval of the utility's governing board.

Design. The consultant's main responsibility is to generate the project drawings and specifications. The owner and consultant agree to various stages of the design, at which times the plans and specifications should be reviewed. The consultant may coordinate the design approval with both local and other regulatory agencies. The consultant usually furnishes an estimate of probable cost of the completed final design. Most of the time, the consultant prepares the bidding documents and construction contract for the bidding procedure.

Bidding. The consultant can assist the owner with the bidding process or conduct the process on behalf of the owner. Specific activities in the bidding process for which the consultant can assist the owner are

Figure 1-4 Interior of a well-supplied project engineer's field office

- recommending publications in which the project should be advertised
- prequalifying bidders (discussed in detail in chapter 4)
- conducting or attending the pre-bid conference
- conducting or attending the bid opening
- tabulating the bidding results
- evaluating the bids and recommending the winning contractor
- preparing a notice of award to be presented by the owner to the contractor

Construction. The consultant may or may not be retained to perform construction contract administration, and the consultant's responsibilities vary accordingly. The following list of responsibilities assumes the consultant is an active participant in the construction phase:

- Perform periodic observation of the constructed project to check for compliance with the intent of the design. The consultant can provide full- or part-time construction observation staff on-site.
- Evaluate construction progress and process progress payment requests for submittal to the owner.
- Provide information and clarification.
- Review shop drawings for compliance with the contract documents.
- Review test reports.
- Conduct and attend preconstruction and construction progress meetings.
- Witness and review tests.
- Report to owner any proposed changes to construction schedule, scope, and budget.
- Review and process change orders for submittal to owner for approval.
- Collect warranty information from the contractor and provide to owner.
- Transpose contractor documented-as-constructed information to record drawings (also called "as-builts").
- Verify that vendors of project components send O&M manuals where required.

- For complex projects, such as a treatment plant, develop project O&M manuals or other documents and provide utility operator instruction.
- Furnish certification of completeness to the owner and applicable regulatory agencies.

Postconstruction. In the traditional construction project, the consultant's role usually ends at the completion of the construction phase. In a few instances, however, the consultant may provide services such as

- assisting the owner with warranty issues
- providing advice on O&M problems
- evaluating and reporting on the project performance after a defined period of time
- assisting the owner in resolving other outstanding issues and litigation

Contractor

Traditionally, the contractor's sole responsibility is to construct the project according to the intent of the plans and specifications. With nontraditional projects (discussed at length in Section II), such as design–build, the contractor's role has been expanded. The following paragraphs focus on the traditional role of the contractor during each phase.

Preliminary and design. In the traditional construction project, the contractor is not involved in the preliminary and design phases.

Bidding. The contractor is responsible for preparing a bid according to the bidding instructions, and may have to submit prequalification documents (refer to chapter 4). The contractor must follow the bidding directions upon receiving clarification of the plans and specifications; such directions are intended to maintain fairness to all bidders.

Along with the bid, the contractor is normally required to submit additional information such as

- bid bond
- questionnaire on ability to perform the project (e.g., past projects, safety plan and record, references)
- list of major suppliers and subcontractors
- preliminary schedule

Construction. The successful bidder enters into a construction contract with the owner. Bound by the contract, plans, and specifications, the contractor is responsible for constructing the project within bid price and within the defined time frame. Construction activities are expected to begin upon receipt of the notice to proceed.

The contractor should follow the lines of communication and authority established in the contract. For example, the contractor should submit progress payment requests as scheduled, give required notice for inspections, attend progress meetings, and negotiate change orders on schedule changes with the owner or contract administrator.

When the project is substantially complete, ready for its intended use, the contractor notifies the owner to demonstrate that the project components have been started, tested, and are performing to the requirements of the contract documents.

If the contract so specifies, the contractor coordinates teaching the owner's staff, often using manufacturers' specialists. In addition, the contractor assembles and gives to the owner the vendors' O&M manuals.

Postconstruction. The contractor's postconstruction responsibilities can be any of the following depending on the contract documents:

- Maintain the bonds active for a defined period of time.
- Maintain completed products' insurance for a defined period of time.
- Return to the site to correct defective work not known at the time of construction completion.
- Coordinate the various manufacturers' return to the site for warranty work.
- In some cases, contractors are engaged to operate the project for a period of time before complete assumption by the owner's staff.

CONSULTANT SELECTION

Although this manual focuses on construction contract administration, a few words on consultant selection may be helpful. The following paragraphs give broad guidelines to owners who are selecting consultants.

Request for Proposal (or Qualifications)

The owner, through direct selection, open advertisement, or a list of prequalified firms, announces a request for proposal (RFP). The RFP describes the project and the intended scope, requests a statement of qualifications (past projects, staff, references), and gives guidelines for submitting proposals. An estimated fee for the work is usually not requested of the consulting firms in the proposal, since at this stage the owner and consultant have not come to a clear understanding of the scope of work the consultant is to perform.

Presentation and Selection

Following review of the qualifications, the owner selects those firms believed to have the capabilities and resources that best fit the needs of the proposed project. These consultants usually make presentations to the owner to demonstrate more clearly their understanding and role in the project. The owner has the opportunity to meet the consultant's proposed team and to ask questions of them.

The owner ranks each of the consultants based on preestablished criteria such as technical ability, understanding of the project, and references. The owner then negotiates a contract with the selected consultant that establishes a detailed scope of services, fee structure, and other conditions, such as maintaining errors-and-omission insurance. Standard contract forms (see chapter 2) for consulting services are available from various organizations.

Fee

The owner and the consultant negotiate a fee after they have both defined in detail the specific project scope. Following arrival at a mutually acceptable fee, both parties enter into a contract. Types of fees are discussed in the following paragraphs.

Lump sum. The consultant and owner agree to a set price, or lump sum, for the agreed scope of work. Obviously, the scope must be clearly defined to make this approach acceptable for both parties. Periodic payments may be made based on percentage of work completed or deliverables made. Any work beyond the original scope may be negotiated in terms of a separate lump sum or on a cost-plus basis with predetermined staff and expense rates.

Cost-plus. In this approach, the consultant and owner usually agree to an upper limit for the agreed scope of work. Periodic payments are based on set fees per hour of staff time (with a markup for profit and overhead) and expenses (often with a markup percentage).

Percentage of construction costs. Less often used than lump sum or cost-plus methods is the consultant fee based on a percentage of the total construction cost. This method is often used as an estimate of fees. The percentage varies according to project size and complexity. This form of compensation is complicated since the design consultant needs to be hired before construction cost is known. Adding to the complexity is the fact that the project may change in scope during any of the various phases, requiring multiple adjustments at the conclusion of the project.

TRADITIONAL PROJECT ADVANTAGES AND DISADVANTAGES

The traditional project has evolved to an extent that it offers some degree of flexibility; however, many new forms of project delivery are available. These alternatives are described in more detail in Section II. In order to discuss the advantages and disadvantages of the traditional project, the main nontraditional alternatives are briefly introduced next.

Nontraditional Alternatives

Other alternatives are available besides the following, but they usually consist of combinations of the traditional form of delivery and alternatives.

Owner-based. In this approach, all phases of the project are performed by the owner if resources and regulations allow. In a sense, this alternative goes back to the prevalent style of many years ago.

Project management firm. The owner hires a project management firm to oversee every phase of the project.

Turnkey. In turnkey and other similar alternatives such as design–build, the owner typically hires one entity (or consortium) possessing the resources of a consultant and a contractor. The firm or firms complete the entire project from preliminary planning to design and construction.

Privatization. When the project scope and budget are defined, the owner can hire a firm capable of performing or gathering the resources to perform all phases of the project, retain ownership, and operate the facility on behalf of the owner. In some cases the facility may be turned over to the owner at a later date.

Advantages of the Traditional Project

Advantages of the traditional project are as follows:

- The owner may not have sufficient staff, expertise, or equipment to undertake portions of the project; the traditional approach makes up for these deficiencies.
- The owner does not need to hire permanent staff for design and construction activities.
- The owner can select the best available experts for each phase of the project.
- Some risk is spread to the other participants.
- Use of outside experts may result in less time needed and lower costs as a result of increased efficiency.

- Use of outside expertise may help convince governing boards of project legitimacy and ability to perform the design and construction.
- All participants are usually familiar with the processes, responsibilities, and expectations.

Disadvantages of the Traditional Project

The disadvantages of the traditional project are as follows:

- As with most other approaches, the responsibilities of participants must be carefully defined.
- The length and complexity of the project may be increased because of the need for multiple outside contracts and contract administration.
- The cost of outside entities needs to be justified to governing boards and shown to be cost-effective.

Each utility brings its own expertise and regulations to the construction process. Some utilities are more flexible than others in determining what form of management the construction project will take. In the end, the owner is ultimately responsible for successful completion of the project.

REFERENCE

F.W. Dodge Division. 1996. *Dodge Reports Bulletin*. 1-800-325-2030. Construction Information Group, New York: McGraw-Hill.

Chapter 2

Construction Documents Review and Application

This chapter discusses the general content of traditional construction documents that define project requirements. Construction documents include the contract, which is the written agreement between owner and contractor. The chapter also reviews standard documents developed by the Engineers Joint Contract Documents Committee (EJCDC). Many water utilities, consultants, and contractors use the EJCDC documents in design and construction projects.

CONSTRUCTION DOCUMENTS AND CONTRACTS

Construction documents must clearly and completely state the responsibilities of the contracting parties. In the traditional construction project, the parties are the owner and the contractor. The elements of typical construction documents (illustrated in Figure 2-1 and Figure 2-2) are (ASCE 1990):

- bidding requirements
- contract forms
- contract conditions (general and supplemental)
- drawings and specifications
- addenda and contract modifications

Bidding requirements include the invitation to bid, the bidding instructions, the project information, and the bid form and bond. The requirements are discussed in detail in chapter 4.

The contract itself is the written agreement that is signed by both parties. Referenced in the contract are other documents that detail the specific project requirements such as the drawings and specifications. The major elements of the contract are

- scope of the work
- responsibilities of the parties (who establishes surveyed grades and horizontal controls, who provides tests of materials, site security, and others)

- effective date of the contract
- definition of terms
- completion dates
- liabilities and indemnification of the parties
- payment procedure and retainage (discussed in chapter 14)
- incentives (refer to chapter 5)
- insurance and bond requirements
- change-order procedure
- settlement of disputes (discussed in chapter 13)
- contract termination procedures
- warranties (discussed in chapter 3)

Legal review of the contract documents by an attorney is recommended, especially for more complex projects.

ENGINEERS JOINT CONTRACT DOCUMENTS COMMITTEE

Various organizations have developed standard construction documents. The main purpose in developing such standard documents is to provide consistency in terminology and procedures. The use of such documents should result in higher quality, quicker completion, and fewer disputes in design and construction. (NOTE: In the following discussion the design consultant is referred to as the "engineer" to be consistent with the documents being described.)

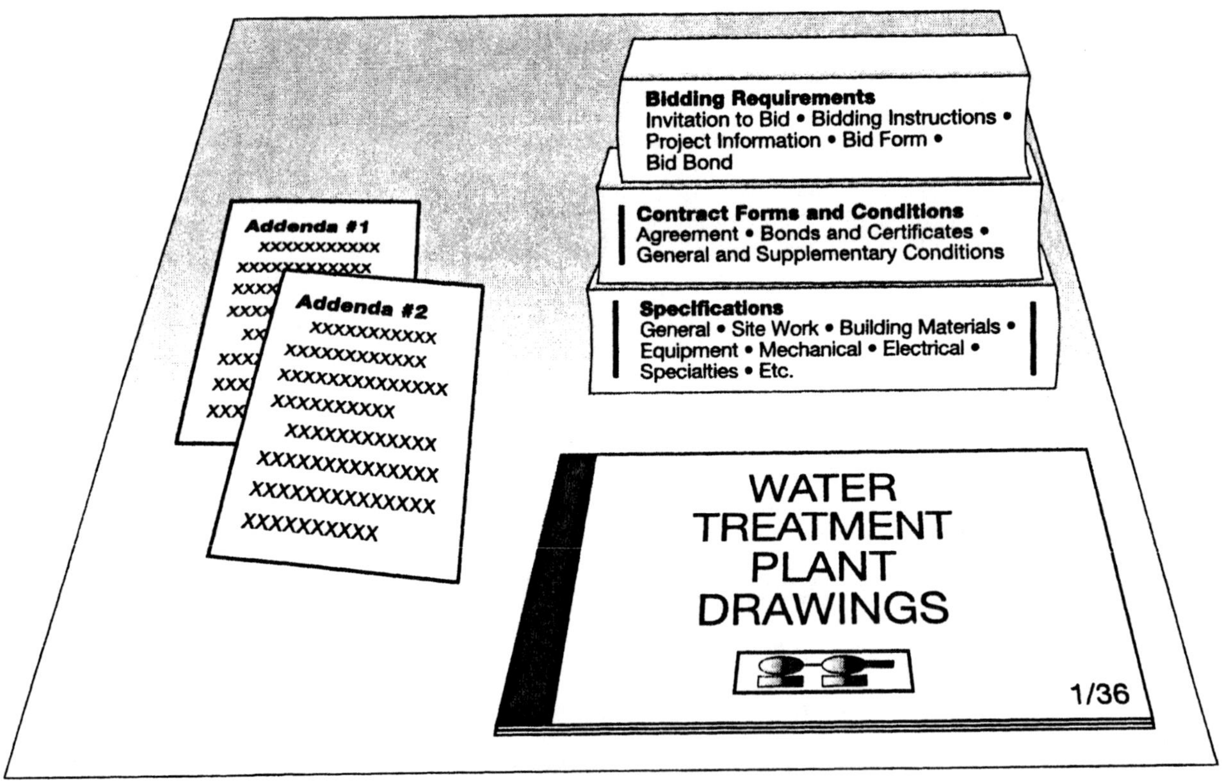

Figure 2-1 Typical contents of bidding documents

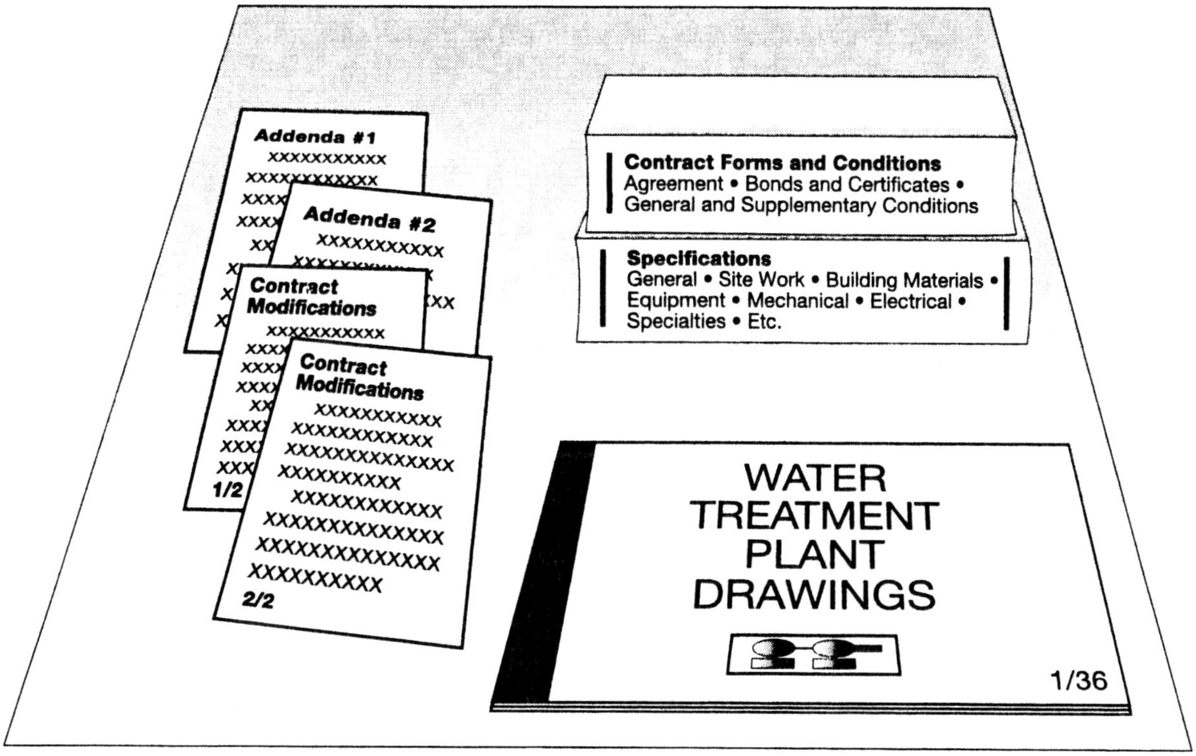

Figure 2-2 Typical contents of contract documents

The Engineers Joint Contract Documents Committee is a national (US) committee of consultant organizations. The EJCDC is dedicated to a long-desired objective of the engineering profession: the development and periodic updating of a set of standard engineering services and construction documents that represent the latest and best thinking regarding contractual relations between the parties involved in a project.

The committee began in 1963 with the formation of a Contract Documents Committee by the Professional Engineers in Private Practice (PEPP), a practice division of the National Society of Professional Engineers (NSPE). The Contract Documents Committee of the American Consulting Engineers Council (ACEC) joined with NSPE-PEPP in 1975. A similar committee of the American Society of Civil Engineers (ASCE) joined in 1978. In 1981, the Construction Specifications Institute (CSI) joined but in 1992 changed its status from member to observer. The overall committee adopted EJCDC as its name.

The committee retains outside legal counsel and in recent years has sought the counsel of other groups in the construction community by establishing liaisons with the American Water Works Association (AWWA), the American Council of Independent Laboratories, the American Institute of Architects (AIA), the Associated General Contractors of America (AGC), and the Association of Engineering Firms Practicing in the Geosciences of Soils and Foundation Engineers.

Through its publications, seminars, and other educational programs, the EJCDC also educates owners, members of the design profession, the construction industry, and attorneys, as well as the interested public, on the importance of understanding the proper roles and responsibilities of the contracting parties and on the wording and coordination of engineering services, bidding, procurement, and construction documents. The standard language and format of EJCDC documents is well known to contractors and can result in more competitive bidding because of fewer ambiguities.

Purpose of EJCDC Documents

The aim of the EJCDC standard documents is to clearly define the functions and responsibilities of the participants in designing and constructing a project. When all participants understand their roles and the roles of the others, the chances for misunderstandings and disputes will be greatly reduced.

The EJCDC standard documents reflect the experience and knowledge of owners, engineers, lawyers, and contractors involved in utility and other types of construction throughout the country. The documents are adaptable to water utilities, large or small, and have been written so that the requirements of a particular owner or project location can be added to the standard text. The standard documents are constantly being updated to reflect changes in project administration.

The EJCDC documents can be divided into four broad categories:

- engineering consulting services
- construction
- procurement (purchasing goods and services)
- commentaries which provide background information and comparisons of different EJCDC documents)

Engineering Services Documents

EJCDC engineering services documents help define standards of engineering practice in the US. Water utilities and engineering consultants use the documents to arrive at acceptable consultant service agreements. Of course, as with any standard document, each specific project must be evaluated for any requirements that might necessitate modifying the standard documents. Procedures for modifying the documents are discussed later in this chapter.

The range of services covered by the EJCDC engineering services documents include the following phases:

- study and report
- design
- bidding or negotiating
- construction
- operation

A list and brief descriptions of EJCDC engineering services documents and form numbers follow:

- 1910-1: Standard Form of Agreement Between Owner & Engineer for Professional Services (long-form agreement includes study through startup and operation)
- 1910-2: Standard Form of Letter Agreement Between Owner & Engineer for Professional Services (short-form agreement includes study through startup and operation)
- 1910-9-F: Guide Sheet for Including Limitation of Liability in the Standard Form of Agreement Between Owner & Engineer for Professional Services (discusses applicability of limitation-of-liability clauses)
- 1910-10: Standard Form of Agreement Between Engineer & Architect for Professional Services (agreement between engineer and subarchitect)

- 1910-13: Standard Form of Agreement Between Engineer & Associate Engineer for Professional Services (agreement between engineer and subengineer)
- 1910-14: Standard Form of Agreement Between Engineer & Consultant for Professional Services (agreement between engineer and a specialized consultant for a limited portion of the design aspects of the project, such as archeology or value engineering)
- 1910-19: Standard Form of Agreement Between Owner & Engineer for Study & Report Professional Services (agreement for study and report only)
- 1910-27-A: Standard Form of Agreement Between Owner & Geotechnical Engineer for Professional Services (agreement between owner and geotechnical subconsultant)
- 1910-27-B: Standard Form of Agreement Between Engineer and Geotechnical Engineer for Professional Services (agreement between engineer and geotechnical subconsultant)

Construction Documents

EJCDC construction documents help define the agreement between the owner and the contractor from bidding, to the contract, to payment requests, to change orders, to the certificate of substantial completion. Again, each specific project must be evaluated for any requirements that might call for modification of the standard documents. Procedures for modifying the documents are discussed later in this chapter as are examples of the potential problems and benefits of standard documents.

A list and brief descriptions of EJCDC construction documents and form numbers follow:

- 1910-8: Standard General Conditions of the Construction Contract (definitions, roles, responsibilities, and procedures to be used during the course of construction)
- 1910-8-A1: Standard Form of Agreement Between Owner & Contractor on the Basis of Stipulated Price (specifies payment method: lump sum, unit price, or a combination thereof)
- 1910-8-A2: Standard Form of Agreement Between Owner & Contractor on the Basis of Cost-Plus (form to specify the cost of work plus fee)
- 1910-8-B: Change Order (for making a change to the contract)
- 1910-8-D: Certificate of Substantial Completion (acknowledgment by all parties that the project has attained the defined substantial completion)
- 1910-8-E: Application for Payment (contractor's application for a progress payment on a multipayment type of contract)
- 1910-8-F: Work Change Directive (issued to a contractor ordering a change in the work and establishing a monetary limit to the change; to be formalized through the use of form 1910-8-B, Change Order)
- 1910-12: Guide to the Preparation of Instructions to Bidders (information to be included in the instructions to bidders)
- 1910-17: Guide to the Preparation of Supplementary Conditions (for use in preparing modifications to 1910-8, the Standard General Conditions of the Construction Contract. Form 1910-9A provides further guidance.)
- 1910-18: Suggested Bid Form & Commentary for Use (a suggested layout and commentary: no single form would be applicable to all types of projects.)
- 1910-22: Notice of Award (establishes the contract price in cases where bids contain more than one project item in the form of alternates)
- 1910-23: Notice to Proceed (establishes start of contract time)

- 1910-28-A: Construction Performance Bond (guaranteeing availability of funds to complete a project in the event the contractor is unable to do so)
- 1910-28-B: Construction Payment Bond (guaranteeing the availability of funds for payment to laborers and material suppliers)
- 1910-28-C: Bid Bond (Penal Sum) Form
- 1910-28-D: Bid Bond (Damages) Form (Forms 28-C and -D guarantee the availability of funds to cover costs in the event the low bidder does not enter into a contract with the owner)

Procurement Documents

EJCDC procurement documents help define the agreement between the owner and suppliers of goods and services. Again, each specific situation must be evaluated for any requirements that might call for modification of the standard documents. Procedures for modifying the documents are discussed later in this chapter, as are examples of the potential problems and benefits of standard documents.

A list and brief descriptions of EJCDC procurement documents and form numbers follow:

- 1910-26-A: Standard Form of Procurement Agreement Between Owner & Contractor (agreement for acquisition of goods and services)
- 1910-26-B: Procurement General Conditions (defines relationships between owner and supplier of goods and services)
- 1910-26-C: Guide to the Preparation of Procurement Supplementary Conditions (describes possible modifications to Procurement General Conditions and additional requirements of alternative procedures)
- 1910-26-D: Instructions to Bidders for Procurement Contracts (instructing bidders on the preparation of their bids)

Commentaries

EJCDC commentaries and updates are descriptions and comparisons of the documents listed in the previous sections. Frequently, several documents are referenced in a single commentary. A list and brief descriptions of EJCDC commentaries and form numbers follow:

- 1910-9: Commentary on Agreements for Engineering Services & Construction Related Documents by John R. Clark. Esq. (origin and use of EJCDC documents)
- 1910-9-A: Commentary on 1990 Editions of Construction Related Documents by John R. Clark, Esq. (commentary and comparison of 1990 version of Standard General Conditions of the Construction Contract to the previous 1983 version; helpful in the preparation of supplementary conditions since it analyzes the reasons for changes from the previous editions)
- 1910-9-B: An Update to Indicate Important Changes in EJCDC Standard Agreements for Engineering Services (presents changes in current documents from previous documents)
- 1910-9-C: Focus on Shop Drawings by John R. Clark, Esq.
- 1910-9-D: Recommended Competitive Bidding Procedures for Construction Projects by Robert J. Smith, Esq. (focuses on creating clear and detailed bidding procedures to facilitate improved competitive bidding documents)
- 1910-9-E: Limitation of Liability in Design Professional Contracts (describes allocation of project risks between owner and engineer)

- 1910-9-G: Indemnification by Engineers—A Warning (discusses applicability of clauses requiring indemnification by engineer)
- 1910-11: Cross Reference Between EJCDC Standard General Conditions of the Construction Contract (1910-8: 1990) and AIA General Conditions of the Contract for Construction (cross-reference to the AIA form of General Conditions; see later section of this chapter discussing development of supplementary conditions)
- 1910-16: Uniform Location of Subject Matter (describes standardized location for the various topics in the bidding and contract documents)
- 1910-20: Engineer's Letter to Owner Requesting Instructions Concerning Bonds & Insurance Regarding Construction (discusses owner's options, project bonding, and insurance limits)
- 1910-21: Owner's Instructions to Engineer Concerning Bonds & Insurance Regarding Construction (conveys owner's bonding and insurance requirements for incorporation into construction documents)
- 1910-25: Advice to Engineers Who Intend Using the 1987 Editions of the AIA's Documents (describes major differences between the AIA and EJCDC documents)
- 1910-26-E: Commentary on Procurement Documents by John R. Clark, Esq. (details origin, terms, and use of the procurement documents)

Modification of EJCDC Documents

Each construction project is unique and EJCDC standard documents cannot cover all situations. Some of the common areas where projects tend to differ are

- specific owner requirements
- local statutory requirements
- type of service to be performed
- the type and complexity of the project
- conditions under which the work is to be performed
- relationships between the parties

Extreme care should be used when modifying or supplementing the standard forms because their provisions are interrelated. Modifications of the language of the standard forms and supplemental provisions should always be made to cover in full the intent of the parties. For example, a change in one place or an addition in another may create an ambiguity or produce an unintended result in the legal obligations of the parties. To reduce such problems, EJCDC documents cross-reference related documents and provide warnings to modify the documents correctly.

Examples

The following examples show how using the EJCDC documents can reduce common problems in construction projects.

Specifying contractor–subcontractor relationships. Potential conflict in construction projects can occur when the owner or engineer attempts to become involved in the relationships between the contractor and the subcontractors. The EJCDC documents are based on the principle that the owner and engineer will deal only with the contractor, not directly with a subcontractor or supplier. The documents provide that the contractor is fully responsible for the acts and omissions of the subcontractors and that nothing contained in the documents creates any contractual relationship between any subcontractor and the owner or engineer.

There are two basic reasons for this approach. One is the strong belief that both the owner and the engineer should avoid undermining a contractor's authority to deal in a businesslike manner with its subcontractors. The other is to avoid discouraging the contractor from fulfilling one of several primary responsibilities: supervising, directing, and inspecting the work and coordinating the activities of subcontractors and suppliers. Several court decisions have held that when owners or engineers undertake the responsibility of supervising and coordinating a contractor's work, they have a duty to perform that function with skill and care and are liable for the results of that action.

On many projects, however, the owner and the engineer desire to retain a degree of control over the contractor's selection of subcontractors. The reason is often an overriding need for expertise in completing critical components of the project. EJCDC documents provide two opportunities for the owner and the engineer to exercise judgment about the capabilities and qualifications of prospective subcontractors and suppliers.

The documents may require that after receipt of bids, the apparent successful bidder obtain the owner's and engineer's acceptance of those subcontractors specifically identified in the documents as being of special importance for the successful completion of the project. If the apparent successful bidder cannot obtain acceptance of a subcontractor nominee before the notice of award is given, that bidder may either furnish an acceptable substitute without an increase in the bid price or withdraw the bid without forfeiting the bid security. If the owner or engineer objects to a subcontractor or supplier after the contract is awarded, the contractor is allowed an appropriate increase in the contract price to reflect any required substitution. The EJCDC believes that it is good practice to resolve the acceptability of subcontractors and suppliers before the contract is awarded and that the allocation of expenses resulting from rejection and substitution before or after award of the contract should be fair and reasonable to all parties.

Clarifying changes to contract price and time. Change orders can be costly surprises for the owner; they disrupt the rhythm and flow of work at the construction site, and they divert the attention of all parties from other concerns equally vital to the success of the construction project. They can also be fertile ground for the seeds of unnecessary conflict. Of course, a perfect set of plans and specifications is not possible; therefore, the owner should establish a contingency fund to cover a reasonable number of unforeseen changes.

A lack of clarity on the issue of changes in the work is seldom the source of conflict. Most contract documents clearly require that if changes are needed that affect cost or time, they can be effected only by a written modification to the contract. The vehicle for such modification is a formal change order reviewed and recommended by the engineer and signed by the owner and contractor.

Serious problems do arise, however, from the way in which change order procedures are followed during construction. Frequently, conditions arise that make the routine processing of a change order impractical. The stakes are high for everyone involved. The pressures of time and cost are often enormous, and no one needs the added pressures of disruption and delay that changes generate.

To provide an improved and workable procedure for handling the change order process, the EJCDC developed the concept of the work change directive. Earlier editions of the EJCDC standard documents provided that changes in contract work, contract price, or contract time could be made only by a change order recommended by the engineer and signed by both the owner and contractor. However, the EJCDC recognized that it is often difficult, if not impossible, for a contractor to estimate, or forward-price, the full impact of contract changes. This difficulty is a primary reason for contractors delaying the execution of change orders—or in some cases even refusing to execute them. To specifically address this problem, the EJCDC developed the concept of the work change directive to permit the contractor to proceed without delay before price and time are finally determined.

When a work change directive form is issued, the contractor must perform the work as recommended by the engineer and as authorized by the owner. This form provides for an estimate of the additional cost and time, which the contractor cannot exceed without prior authorization. But because the contractor is not required to sign the work change directive form, its right to request additional relief, if required, is reserved. When the owner, contractor, and engineer agree on the total effect of the change in the work, a change order properly signed by all parties is issued. Although the work change directive is relatively new, those who have used it report it has significantly reduced conflicts during construction.

Another area of conflict in the change-order procedure is the basis for pricing changes in the work. Most standard documents provide that the value of the work covered by a change order will be determined in the following order: (1) by the application of unit prices previously bid, (2) by a mutually agreed upon lump sum, or (3) on the basis of cost of the work plus a fee for overhead and profit. Because of the need to justify the cost of changes to higher authority or to regulatory or grant agencies, many owners require that changes be performed on the basis of the cost of the work. However, although many owner's documents state the allowances that can be used for overhead and profit, the documents frequently are silent, or sketchy at best, on what charges do or do not constitute "cost of the work" and to what "costs" the allowances for overhead and profit apply.

Clear and unambiguous definitions of these terms can prevent unnecessary conflict. The EJCDC documents were developed for use with contracts based on a unit price, lump sum, or cost plus fixed fee. As such, these documents contain comprehensive definitions of "cost of the work" and contain guidelines on applying allowances for overhead and profit. The definitions are useful in that they provide a "road map" for the forward-pricing of a change order when changes are to be performed on the basis of a mutually agreed upon lump sum.

Lessening delay and acceleration problems. Another way conflicts can be readily minimized is to avoid using unenforceable clauses in documents. Such clauses make contract administration more complex and may prompt the owner to resist settling legitimate claims.

A specific example of this is damages for delay. Delay in construction projects is so common that completing a project on time today is considered unusual. Contractors are almost invariably subject to the proposition that "time is money." Delay in performing the work or acceleration to overcome the delay frequently results in cost increases for the contractor that far exceed the direct cost of the change or of the changed condition that caused the delay. For these reasons, delay damages have become a major factor in many litigated or arbitrated construction claims. Careful evaluation of the project schedule is the basis for determining responsibility and extent of an excusable delay. There may be a concurrent delay over which the contractor has control, in which case damages are adjusted accordingly.

When the contractor causes the delay, the contract documents are usually clear and stipulate an amount in the form of liquidated damages to compensate the owner. But what about other types of delays, such as those caused by the owner or by circumstances beyond the contractor's control? For example, what should be done when a severe flood washes out the backfill under a treatment plant, as shown in

Figures 2-3, 2-4, and 2-5? When the contract documents do not address delays beyond the contractor's control, the courts usually hold that the contractor is entitled to recover damages.

Historically, many contract documents have contained "no damage for delay" clauses that preclude monetary adjustment and attempt to limit the contractor's recovery to an extension of time. Interpreted literally, these clauses may produce a harsh, unfair result. As a consequence, many courts have interpreted these clauses narrowly, frequently finding that the contractor should be compensated for a delay in situations in which the clause's literal reading would have barred recovery. The courts' reluctance to enforce "no damage for delay" clauses arises from the inequity that such clauses engender.

The EJCDC believes the prudent course of action is to provide in the contract documents for monetary adjustment when a delay occurs. The documents provide that (1) the contractor may be entitled to delay damages when the owner causes the delay and (2) the owner is entitled to delay damages (in the form of liquidated damages) when the contractor causes the delay. Additionally, provisions are made for an extension of time following delays not caused by either party.

Allowing for unanticipated conditions. All contractors who perform work under fixed price contracts determine their bids on the basis of certain fundamental assumptions concerning the conditions they will encounter and the effect of those conditions on price and time. Unanticipated conditions such as concealed subsurface conditions or concealed physical conditions in or around existing structures can greatly disrupt project scheduling and otherwise increase the cost of performing the work.

The owner has an affirmative duty to disclose to a contractor information about conditions that will affect cost, progress, or performance of the work. This information usually includes reports of exploration and tests of subsurface conditions that the engineer has used to prepare the contract documents. If the work is to be performed

Figure 2-3 Severe washout damage at a treatment plant

Figure 2-4 Stabilization of creek banks with rock-filled gabions

in or around existing structures, the information usually includes the owner's "record" drawings of those structures.

In an attempt to discourage contractors from inflating their bid prices to cover the contingency of unanticipated conditions, most contract documents include a "changed conditions" or "differing site conditions" clause that allows price or time or both to be adjusted when such conditions are encountered. On the other hand, to reduce the risk assumed under such clauses, these documents also include disclaimers concerning the accuracy or adequacy of the information provided and a related "site investigation" clause that requires contractors to make, at their own expense, all such investigations and tests as they determine necessary to prepare their bid.

Reliance on the "changed conditions" clauses and associated disclaimers, particularly with respect to subsurface conditions, has not had the desired effect. Contractors continue to include significant contingencies in their bids to provide for unanticipated conditions. Additionally, contractors' lawyers have been successful in the courts in showing the inequity of a design–construction system that puts virtually all of the responsibility for subsurface conditions on contractors and does not provide adequate time or funding for them to identify those conditions properly. Stronger wording of disclaimers has not been an effective method for changing these results. As a result, a distinct trend requires the owner to bear the costs when information about the subsurface does not describe the actual conditions.

In recognition of these trends, the last two editions of the EJCDC standard construction documents take a somewhat different approach to the sharing of risks by the owner and contractor for unanticipated subsurface or physical conditions. As previously discussed, the documents provide for furnishing the contractor with all information on subsurface and physical conditions used by the engineer in preparing the contract documents. The significant change is in the disclaimer for the furnished information. The documents now provide that the contractor is entitled to rely on the

Figure 2-5 Equipment used in creek bank repair

technical data contained in the furnished information but not on nontechnical data, interpretations, or opinions. With regard to subsurface conditions, the "technical data" would include the boring method, location of borings, identification of materials encountered, level of subsurface water, laboratory test methods and results, and similar factual data, all as of the date(s) made.

The EJCDC believes this more equitable basis for sharing the risks for unanticipated subsurface or physical conditions will encourage contractors to decrease the contingency amounts in their bids and will reduce conflict by providing a better basis for determining an unanticipated or changed condition.

REFERENCES

American Society of Civil Engineers (ASCE). 1990. Quality in the Constructed Project: A *Guide for Owners, Designers, and Constructors*. Manuals and Reports on Engineering Practice No. 73. New York: ASCE.

Engineers Joint Contract Documents Committee (EJCDC). Documents, forms, commentaries, and other publications. Washington, D.C.: American Consulting Engineers Council; New York: American Society of Civil Engineers; Alexandria, Va.: National Society of Professional Engineers.

Chapter 3

Warranties

Warranties guarantee the materials and work supplied by the contractor and product suppliers for a prescribed use and length of time. The owner should recognize that more stringent warranties often result in higher costs. As a simple example, the only differences between water heaters having 5-year and 10-year warranties may be the length of the warranty and the price of the water heater. There is often no difference in construction or materials.

The main warranty issues discussed in this chapter are

- participant responsibilities
- general content
- conditions and requirements
- claims

PARTICIPANT RESPONSIBILITIES

The primary participants in a project—owner; consultant (engineer or architect); contractor; and the manufacturers, suppliers, and vendors—have responsibilities related to warranties. These responsibilities are introduced in the following paragraphs; most are also discussed in more depth in later sections of the chapter. The responsibilities described are those common in a traditional project, and they will vary from project to project.

Owner

The owner is responsible, often with assistance of the consultant, for defining the desired warranty requirements of the facility to be constructed or materials to be supplied. These requirements include the duration of the warranties as well as any special warranty requirements such as warranty bonds and factory representative maintenance and inspections.

After completion and acceptance, the warranty period begins. To keep a warranty in effect, the owner must correctly operate and maintain the facility and equipment according to the manufacturers' requirements. In addition, the owner has to follow correct notification procedures when potential warranty-covered problems are encountered. Portions of a project may go into operation in phases. The warranty period for each phase usually begins when the owner takes over the phase, depending on specific contract language.

Design Engineer

The owner's engineer or a consultant specifies the warranty requirements in the contract documents. During construction, the engineer or construction manager verifies that the equipment and material are in proper condition when received after shipment, are properly stored and protected during construction, and are properly installed and started up by the contractor. These conditions are necessary to prevent voiding any warranties. After construction is complete, the consultant may continue to advise the owner on warranty issues.

Contractor

The contractor usually provides a warranty on all construction which it has performed or subcontracted. The contractor is responsible for purchasing, receiving, storing, installing, and starting up equipment and materials from the suppliers according to the construction specifications and the suppliers' requirements. The contractor performs all work correctly to ensure that all warranties are maintained in full force and effect. The contractor includes all warranty provisions in its purchase agreements with subcontractors and vendors and obtains appropriate written warranties to be provided to the owner. The contractor provides all maintenance until the owner officially assumes that responsibility.

Manufacturer or Supplier

The manufacturer or supplier must manufacture, fabricate, and assemble the equipment and materials according to the specifications and is usually responsible for proper shipment to the point of use. The manufacturer or supplier defines all manufacturer warranty requirements that include shipping, installation, and maintenance. When required by the owner, the manufacturer or supplier may assist or supervise installation, startup, testing, and maintenance. For example, Figure 3-1 shows the installation of nanofiltration membranes. The manufacturer or supplier also responds to problems with the equipment or materials according to the warranty agreement.

GENERAL CONTENT

Warranties are of two general categories. The first is a warranty on a specified item of equipment or material by a supplier, vendor, or manufacturer directly to the owner. The second is a warranty of a project or facility by a construction contractor. In the warranties of a project or facility, the construction contractor guarantees its work and materials. When the owner assumes control of the project, any equipment and material warranties provided by subcontractors, suppliers, vendors, or manufacturers then become the owner's responsibility. If a warranty-covered problem occurs, a question often arises as to whether it is a manufacturer's defect, a contractor's installation problem, or an owner's operations and maintenance (O&M) problem.

The warranty conditions or requirements must be established by the design engineer when the specifications are developed. These requirements must be clearly

Figure 3-1 Installation at 12-mgd (0.5258-m^3/s) nanofiltration water treatment plant in Collier County, Fla.

and fully described in the specifications and included in the purchase order or bid documents. Warranty specifications include three primary elements:

- content and extent of the coverage
- duration of the warranty
- effective date when the warranty starts

Equipment and Materials Warranties

The warranty of equipment and materials guarantees against failure of the equipment or materials to perform or operate as specified in the purchase agreement or construction documents. The warranty covers improper design, defects in the materials, and defects in the fabrication, manufacture, and assembly. If the manufacturer does not install the equipment, the warranty usually excludes the manufacturer from liability caused by incorrect installation. Installation procedures should be documented by the owner in case questions arise. Photographs provide a good record of installation (Figures 3-2 and 3-3).

Although a typical warranty period is one year, the duration may vary considerably depending on the type of equipment and materials. For example, some warranties may be 90 days for electromechanical equipment to several years for pipe, roofing materials, or pond liners. The owner may require that a contractor or manufacturer obtain a bond for long-duration warranties. However, warranty bonds are expensive and should generally be secured only for very costly items with long warranty periods. The bonds protect the owner in cases when a contractor, supplier, or manufacturer is no longer available to respond to a warranty problem.

The effective date of the warranty may be upon receipt of the equipment by the owner or contractor, upon installation and acceptance of the component, or upon substantial completion of the entire project. The choice depends on the type of equipment and project.

Equipment or material warranty specifications should specify (1) the manner in which the supplier, vendor, or manufacturer is to respond to the owner when

Figure 3-2 Installation of tee for a 60-in. (1,500-mm) water main

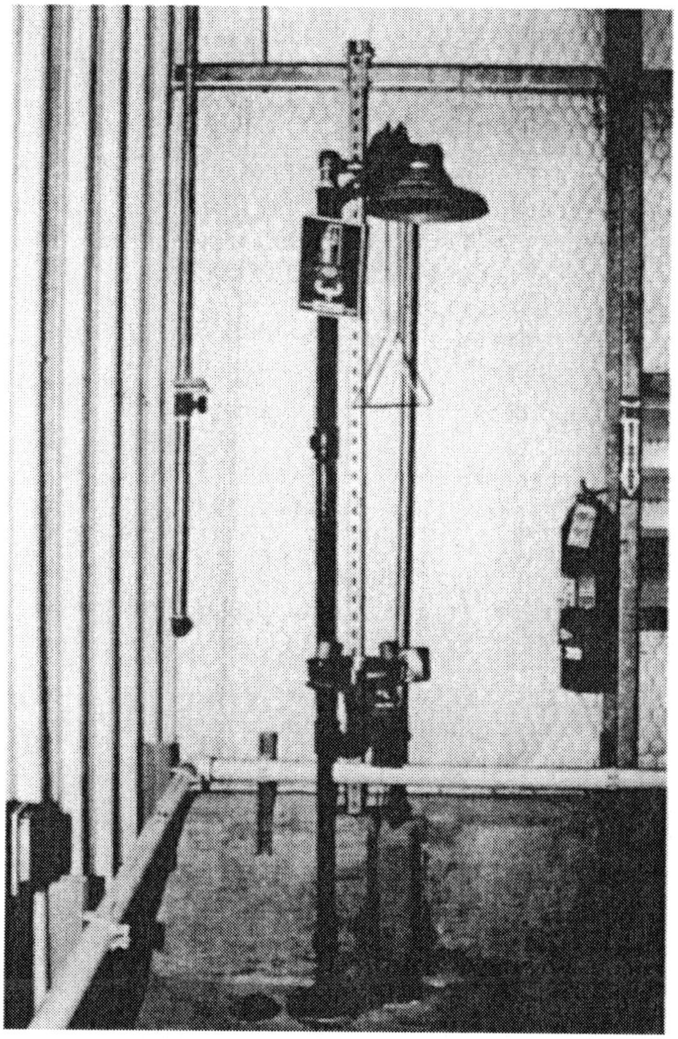

Figure 3-3 Chlorine decontamination equipment

notification of a potential warranty problem is given by the owner and (2) the length of time given the supplier, vendor, or manufacturer to respond to the owner's notification.

Projects or Facilities

Warranties of projects or facilities are often more complex than equipment or material warranties—more participants are involved, and the variations of responsibilities increase. Warranties must clearly establish who is responsible for the design, construction, and operation of the facility. Additionally, a contractor must require its subcontractors, suppliers, vendors, and manufacturers to guarantee their work, materials, equipment, or systems to the owner.

The duration of a facility warranty is typically one year or longer. The date the warranty becomes effective is usually the time of substantial completion. As is the case with equipment or material warranties, the specifications for a project or facility warranty should specify (1) the amount of time the contractor has to respond to the owner when notification is given by the owner of a potential warranty problem and (2) the manner in which the contractor is to respond to the owner.

Nontraditional Construction Contracts

In nontraditional owner–contractor relationships, the contractor often assumes *additional* responsibilities; these will be covered in detail in later chapters. As a brief example, in design–build projects, the design–build contractor must guarantee its project designs or designs by its subcontractor.

CONDITIONS AND REQUIREMENTS

After a purchase order or contract is signed, several activities occur that affect the warranty. These activities include

- preparation of shop drawings
- receipt and storage of equipment
- installation
- operation and maintenance

Shop Drawings

During the initial phases of construction and fabrication of equipment, the supplier, vendors, and manufacturers of the equipment and materials prepare shop drawings. Basically, shop drawings detail the equipment and materials to be provided, for comparison to the specifications.

Construction documents normally dictate that warranty conditions and requirements must be included in the shop drawing submittal. Such conditions and requirements include shipping, unloading, long- and short-term storage, installation, startup and testing requirements, maintenance requirements, and requirements for extended nonuse by the owner ("mothball requirements").

Shop drawings should also state whether a factory representative must be present during the installation, startup, or testing of the equipment or material. Failure to have the factory representative present may void the warranty. For owner-furnished equipment, the owner informs the contractor whether a factory representative must be present.

Receipt and Storage

When equipment or materials arrive at a project, they should be thoroughly examined by the contractor and the owner's representative. Shop drawings and warranty information should be used to determine that the proper materials or equipment were shipped; there was no damage in shipment; and there are no readily apparent defects, damage, or problems that could affect the ultimate use of the equipment or materials. The initial inspection should be documented and filed for possible future use if a potential warranty problem occurs.

Shop drawing information should specify how to store the equipment and materials adequately. Common storage concerns are

- heat
- cold
- precipitation
- sunlight
- wind-blown dust
- sandblasting

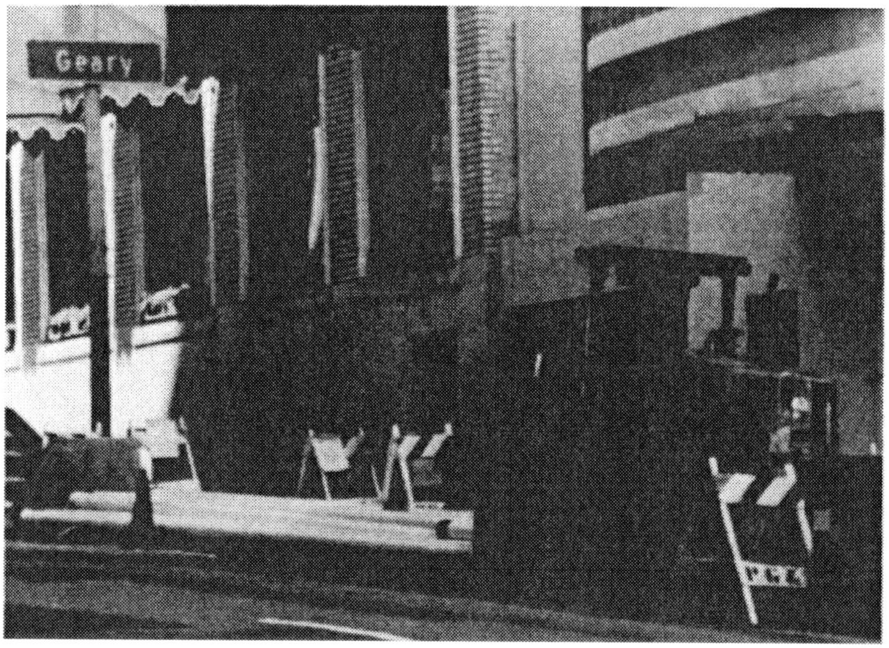

Figure 3-4 Pipe stored on a busy street

- paint overspray
- physical damage
- vandalism or theft

When equipment and materials are in storage, they should be inspected periodically to verify that proper storage is maintained. Periodic maintenance activities on stored equipment are often part of the manufacturer's requirements. If there are potential problems during storage (such as pipe stored on a busy street, as shown in Figure 3-4), the contractor should be informed in writing of the potential problems and be given specific directions on correcting the deficiencies. An evaluation should also be made as to whether to provide written notification to, and require a response by, the manufacturer.

Installation

Proper installation of the equipment or materials must be part of the manufacturer's requirements, which are normally included as part of the shop drawings. All affected parties should have a copy of the installation requirements.

Some of the concerns during installation include

- handling (Figure 3-5)
- placement (Figure 3-6)
- connections (Figure 3-7)
- backfilling (Figure 3-8)

After installation and before the owner accepts equipment from the contractor, each item of equipment must be thoroughly tested to verify that it meets the requirements of the specifications, was properly fabricated by the manufacturer, and was properly installed by the contractor. Each item of equipment has a unique set of criteria that must be verified during the test, and acceptability should never be assumed. A complete report for each test should be made by the engineer and each area of compliance and deficiency recorded. This record is important because it will

Figure 3-5 Handling a 60-in. (1,500-mm) waterline

provide documentation necessary to verify problems or show that startup and testing were performed according to the manufacturer's requirements.

On complex systems or with costly equipment, a manufacturer's representative is usually present to inspect, assist, or direct the startup efforts. This person ensures that the manufacturer's requirements are fulfilled and provides highly competent technical expertise to the project during startup. The presence of a manufacturer's representative also helps reduce warranty-related problems after operations have started.

On complex projects a formal startup and test plan is developed to verify that the total facility meets the specification requirements. The test plan should be developed and executed in a manner to ensure that all equipment and material warranty requirements are met. Test plans should also be developed to show that

Figure 3-6 Placing an 80-in. (2,000-mm) diameter steel pipeline in suburban Washington, D.C.

each specification requirement is fulfilled and that the work, as part of an integrated facility, conforms with contract documents and meets warranty requirements.

After equipment is installed, started, and tested, and the project successfully completed, the owner issues a letter of acceptance of substantial or final completion. After the acceptance letter is issued, the responsibility for proper maintenance and operation passes from the contractor to the owner. The date of the letter of acceptance is normally the date the warranty period begins.

Operations and Maintenance

Improper maintenance and operations represent the most frequent cause of voiding a warranty. To keep warranties in effect, the owner must operate and maintain equipment and materials according to the manufacturers' requirements. To ensure that this occurs, the owner must take three primary steps:

- The owner receives and maintains a functional library of all manufacturers' O&M instructions.
- The owner develops and implements facility-wide O&M programs including operator training manuals.
- The owner performs initial and periodic operator training.

The manufacturers' O&M instructions should be received during construction and placed in a formal library system. The library system should include a master file index for use by the O&M supervisory and engineering staff and for making additional copies when the operations copies are lost or worn. The library system should also include second working copies for use by the O&M staff.

The manufacturers' O&M instructions are used to develop an integrated owner's O&M program and operator training program. These programs produce an O&M manual and an operator training manual. Periodically, a manufacturer's representative is required to provide operator training. These sessions can be videotaped for future reference.

The owner must closely adhere to the O&M instructions issued by the manufacturer to keep the warranty in effect. The owner must properly implement a programmed O&M program throughout the life of the equipment, especially during the warranty period. It is this preventive maintenance program that provides the

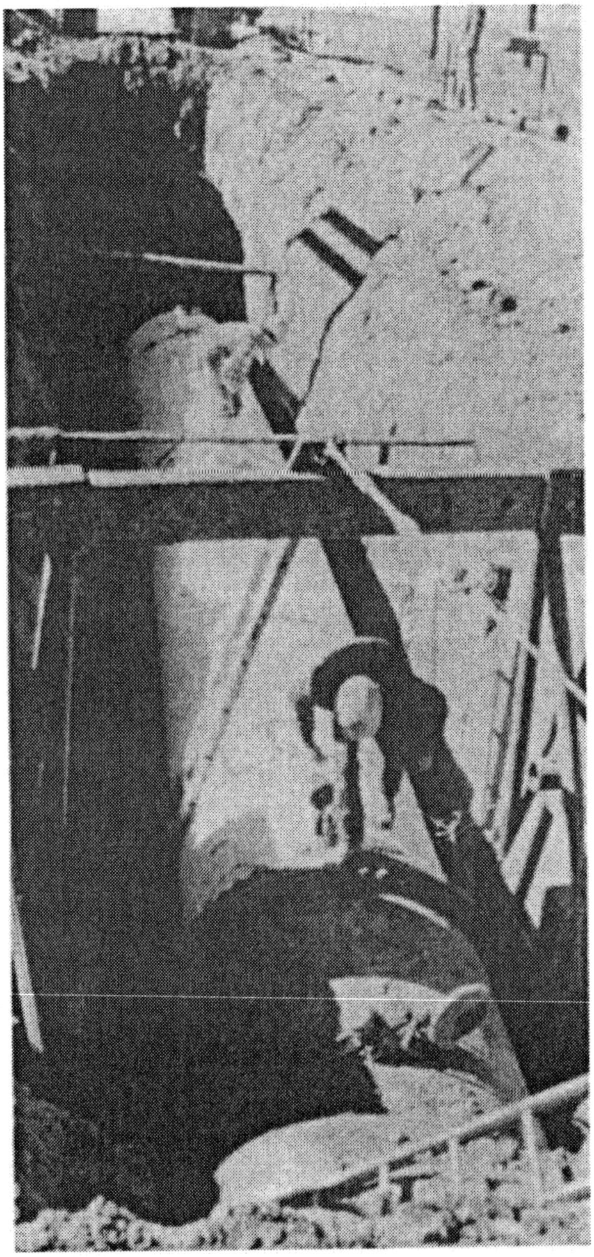

Figure 3-7 Sixty-in. (1,500-mm) waterline connections

Figure 3-8 Sixty-in. (1,500-mm) waterline backfill

owner's legal protection under the warranty and a maximum useful life for the equipment or project. Figure 3-9 illustrates a preventive maintenance procedure.

CLAIMS

If an apparent warranty problem occurs, the owner must inform the contractor both verbally (over the telephone or in person) and in writing immediately; in some cases the owner should inform the manufacturer too, especially if the contractor does not respond adequately. The letter should provide a description of the problem, the date the problem occurred, and a reasonable time for the contractor and manufacturer to respond. If supporting documentation is available, it should also be provided to the contractor and manufacturer. Use of certified mail is advisable.

The owner should provide documentation of potential warranty problems (1) to resolve the warranty problem adequately and as quickly as possible and (2) to

Figure 3-9 Operator cleaning the socks of a bag-type dust collector installed to prevent loss of lime dust to the atmosphere

determine who is responsible to correct the problem. The owner must be prepared to make a fair and reasonable payment if one party responds to a potential warranty problem and it is determined that another party is actually responsible for the problem.

If the contractor does not adequately respond to the problem, then the owner must inform the manufacturer (if applicable) to correct the problem. If neither the contractor nor the manufacturer is responsive, then the owner may be required to correct the warranty problem and take subsequent legal action to evoke the warranty claim.

Chapter 4

Prequalifying and Selecting Contractors

This chapter focuses on selecting qualified contractors for construction projects. Generally called the competitive bidding phase of a project, specific contractor selection procedures vary among states, entities, and projects. This chapter introduces the principal steps in a typical bidding process, including the steps in prequalifying contractors.

Prequalification is the selection of contractors who will be allowed to bid on a project. Not every project involves prequalifying contractors; indeed, regulations may not allow some entities to use prequalification.

The chapter begins with a discussion of prequalification, including its advantages and disadvantages. The remainder of the chapter presents the process of competitive bidding.

PREQUALIFICATION

Whether the contractor will be selected through competitive bidding with open advertising or through a limited bidding of a list of contractors, the contractor who eventually wins the project must be able to perform the work. The owner must make sure that the winning bidder is qualified. The difference is that with prequalification, only those contractors considered to be qualified are allowed to bid.

The following discussion presents the general issues, advantages, and disadvantages of prequalification, as well as the steps of the process.

General Issues

The general issues involved in prequalification are discussed in the following paragraphs.

Goals. The goal of prequalification is not to select the most desirable bidders, but to eliminate unacceptable bidders. The important goals of prequalification are

- to increase the probability of a successful project by selecting a contractor that has the appropriate experience, a strong financial position, and proper staff and equipment
- to reduce the possibility of delays, cost overruns, and associated lawsuits
- to obtain equipment or materials that are compatible, have a proven record of performance, and are from a manufacturer who will provide necessary service

In itself, prequalification does not mean the project will be of high quality; nor will it reduce the need for complete and accurate plans and specifications or reduce the need for adequate construction administration.

Regulations. For most general construction work, public entities are bound by traditional advertised bidding procedures. Often the only contractor qualifications required are valid licenses to perform the work and financial capability to enter into a contract by providing bid and performance bonds (ASCE 1990). Privately owned entities, on the other hand, may be able to skip the competitive bid process and contract with firms with whom they want to work. For projects that are more specialized or may have unusual characteristics (Figure 4-1), such as an environmentally sensitive area (Figure 4-2), the regulations under which public entities operate may allow for prequalification.

Local laws and regulations might also impose specific criteria for prequalification such as requirements for disadvantaged business participation, use of union-made products, disqualification of products made in specific foreign countries, and requirements for US-made products or components. Because of the variance of

Figure 4-1 Tunnel for a 96-in. (2,400-mm) diameter steel pipeline in suburban Washington, D.C.

Figure 4-2 Seventy-mi (112.6-km) pipeline consisting of 54- and 30-in. (1,350- and 750-mm) diameter pipe, Yucca Valley, Calif.

regulations, the owner or engineer should verify that prequalification is allowed (or possibly required).

Disqualification of a prequalified bidder. When a firm has been prequalified, it may be very difficult to disqualify that firm after bids are received and opened. Prequalification documents should make clear that (1) all bidders must meet the qualification requirements detailed in bidding documents and (2) prequalified bidders could be disqualified if those requirements are not met.

Other examples. There are other contracting situations besides the selection of prime contractors that might benefit from prequalification. Some common examples are

- establishing a standing list of firms eligible for certain kinds of projects (such as pipeline repair, as shown in Figure 4-3), rather than just for a single project
- establishing a list of subcontractors or materials suppliers, including specialty subcontractors (e.g., instrumentation, supervisory control and data acquisition or SCADA systems, laboratory equipment and supplies)
- continuing consultant services
- continuing services for construction projects of a routine nature (e.g., short pipeline installations, rehabilitations, or relocations; package lift stations and short force main construction) based on established or negotiated unit prices
- establishing lists of materials suppliers (meters, valves, hydrants)

Standing lists of prequalified contractors, subcontractors, and suppliers are usually valid for a specific time period, such as a year or two. Before the time period

Figure 4-3 Full-circle repair clamp on older waterline

ends, a call for firms to submit prequalification documents is issued. Contractors may be more amenable to this process if the list is compiled fairly, if it is updated routinely, if there is a substantial amount of work to be awarded, and if the work is distributed among eligible contractors.

Another similar practice by design consultants is the naming of acceptable manufacturers of specific equipment in contract documents. The list is based on the requirements of the projects and the consultant's knowledge of products that meet them.

Advantages

The advantages and disadvantages of prequalification vary with every project and must be evaluated to determine whether prequalification is justified. The advantages of using prequalification for a particular project depend on such things as location, complexity, timing, local bidding climate, and owner's and design consultant's experience with that particular type of project.

Advantages include benefits to bidders and owner. The owner benefits from accomplishing the goals of prequalification stated previously in this chapter. The owner reduces the time required to review and evaluate bids from unqualified bidders. It is easier for the owner to disqualify a bidder before bids are received than after they are opened.

Bidders and owners benefit from elimination of unqualified competition that could skew the bidding process by submitting unqualified low bids. Owners may save time over a typical full contractor selection process by using a standing list of acceptable contractors.

Bidders benefit from advanced knowledge of project requirements that could disqualify them; thus, unqualified bidders avoid the expense of preparing a bid and save themselves from the possible embarrassment or aggravation of public disqualification. A bidder's thorough, objective review of project requirements and its own qualifications could allow it to avoid the difficulties or losses that would result from contracting for work for which it is not qualified.

Continuing service agreements are typically used to eliminate or shorten the time-consuming selection process. The agreements are often used when a number of projects are anticipated and repetition of the normal selection process would be burdensome.

Owners may save on project costs by assuming the risk and eliminating or reducing the need for surety bonds from prequalified contractors.

Disadvantages

Most disadvantages associated with prequalification concern criteria that do not accurately evaluate a contractor's ability to complete the work successfully. For example, an owner who may have had a bad experience with a contractor who used rental equipment may wish to eliminate contractors who use rental equipment.

Prequalification may be viewed as a subversion of the general competitive bidding procedures. Prospective bidders may be disqualified based on some criteria that could be arbitrary, contrived, or based on a purely speculative concern for avoiding potential project difficulties. In these cases, the owner may be perceived as favoring a few contractors.

Prequalification may increase project costs by eliminating competition among bidders or by eliminating bidders who might have an innovative and cost-saving approach to executing the work. Disqualified bidders may be stifled in their growth if they are eliminated from projects in which they do not have experience even though they may be able to perform adequately.

Other examples of disadvantages to potential bidders would be cases where bidders are eliminated (perhaps because of poor performance on a previous project that is not representative of their current staff and ability. Examples include performance by subcontractors, circumstances beyond the contractor's control, or ill-advised actions of the owner).

Prequalification Procedure

The following paragraphs identify the steps the owner (or consultant or both) should take to prequalify prospective bidders for a project. These steps are presented in approximate order of completion, although some may occur concurrently. Usually the project design is complete or nearly so when the process begins.

Determine advantages and disadvantages. Review the project design and conditions to determine whether prequalification is desirable for the project or certain of its components. Weigh advantages and disadvantages as described earlier in this chapter. If the advantages clearly outweigh the disadvantages, then proceed with prequalification.

Identify critical characteristics. Identify the critical characteristics of potential bidders as required for the project. Set minimum acceptable criteria below which potential bidders would be disqualified. Depending on the project, the types of information to be submitted by the contractor include

- performance and experience records
- experience with similar work
- types of materials or equipment produced
- financial status (ownership)
- equipment owned

- personnel on staff
- pending litigation
- bonding capacity
- minority or women's business enterprise status

Note that there may be potential problems with requesting certain kinds of information (financial data, pending lawsuits) that by law are confidential. An attorney should review the types of information requested. Sometimes alternative information could be acceptable; for example, rather than providing a financial statement, a letter from a bonding agency might be acceptable. (NOTE: The contractor may also desire to learn the financial status of the owner.)

Prepare and send the statement-of-qualification form. Standard forms can reduce the time and expense of preparing prequalification forms and the time reviewing responses. One such form is Associated General Contractors of America (AGC) Document No. 220, *Construction Contractor's Qualification Statement for Engineered Construction*. The form should be evaluated to verify that it fits project conditions and owner requirements. Whichever form is used, it should clearly and precisely state which information is required; this will minimize the need for clarification or additional data.

The next step is to announce, advertise, publish requests for statements of qualifications (RFQs) from prospective bidders, and send RFQs directly to known prospective bidders. Include detailed instructions and explain precisely the minimum qualification requirements. In addition, specify when prequalification decisions will be made and whether disqualified firms will be notified of the basis for decisions. If appeals of prequalification decisions are allowed, the associated procedures should be detailed.

Projects may require subcontractor prequalification as well. The RFQ should be sent out far enough in advance so that prospective bidders do not expend time and effort contacting subcontractors that are not prequalified.

Evaluate the qualification statements. The owner or engineer usually establishes a selection team of three to seven members to evaluate the statements of qualifications. The responsibilities and authority of the selection team—that is, whether the team recommends qualified applicants to the decision maker or makes final decisions—should be clear. Team members must understand selection criteria. When qualified bidders are selected, the list should be sent to all applicants.

COMPETITIVE BIDDING PROCESS

The intent of the competitive bidding process is to select a qualified contractor at the most competitive price, which usually is the lowest bid. For this process to work, the owner must prepare clear, complete, and detailed plans and specifications and also clearly state the necessary qualifications of the contractor. Bidding instructions and requirements must also be clear.

The remainder of this chapter presents the main steps and documents used in the competitive bidding process. This section assumes the project is publicly funded; privately funded projects can use the procedure as well, although they may not be bound by law to do so.

Documents

Bidding documents are strictly used for advertising the project and giving instructions to the bidders. Restatement or explanation of contract or technical provisions must be avoided (Smith 1987). Standard Engineers Joint Contract

Documents Committee (EJCDC) bidding documents were listed and briefly described in chapter 2. Typical bidding documents are described in the following paragraphs.

Advertisement and invitation to bid. The advertisement and invitation to bid contains the owner's name, a brief description of the project, bidding schedule, and an explanation of where to obtain the bidding documents. The advertisement and invitation is called the legal notice and is sent to contractors and placed in appropriate newspapers and periodicals. If a monetary deposit is required for the bidding documents (which includes a complete set of plans and specifications), the amount should be specified. Plans may also be available for viewing, but not removing or copying, in the owner or consultant's office or other agency's "plan room."

Instructions to bidders. The instructions to bidders detail the steps necessary to submit a bid and vary from project to project. Instructions generally include documents that must be completed, information that must be given, deadline for bid submittals, bid-opening procedures, and contract-awarding procedures. If a prebid conference is scheduled, the date and time should be included in the instructions. Instructions should also suggest strongly that bidders visit the project site.

Bid form. The bid form lists the line items of the project and the bases for the bid (e.g., cubic yards, lineal feet, lump sum). The bases should be standard measurements used in the industry. The form also has spaces to list totals for each item and the grand total of the bid.

Bidder qualification. This information is required by the owner to ensure that bidders are qualified and able to complete the project. A list of subcontractors and their qualifications may also be required. Typical qualification information was discussed previously in this chapter.

Bid security and bonds. The bonding amount required is usually 5 percent of the bid, though it may be higher. A bid bond form is usually included in the bidding documents, although most bond agencies use their own forms.

Addenda. During the course of bidding, bidders may raise issues that need clarification. If one contractor receives such clarification, or if errors are found that need to be corrected, all bidders must receive the clarifications or corrections. Official addenda are sent to bidders, who must acknowledge receipt, usually by signing the addenda as received and submitting them with the bid.

Bid Opening, Evaluation, and Award

The following procedures are performed by the owner or consultant on behalf of the owner.

Bid opening. The time to open bids must strictly be observed; late bids should be treated as if they were never received (Smith 1987). Bids normally are randomly selected and opened and the grand total read aloud. A bid tab form listing bidders is filled out with the prices; bidders can be given a copy of the blank form to fill out as bids are read. An official bid tab should be sent to all bidders.

The receipt of addenda and other required forms should also be announced at the opening. The original bid forms and documents may be examined by bidders in the presence of the owner or consultant.

Bid evaluation. Assuming the low bidder has complied with bidding instructions (especially that the bid was signed), the next step is to verify the bid. If it seems too low, considerably less than the next lowest bid, the bidder should be asked to verify the bid. If the bidder has made a verifiable mistake, the bid may be withdrawn. Forcing the bidder to perform at the quoted bid is rarely to the owner's advantage. If the low bid is valid, the bidder's qualifications are verified according to preestablished criteria. The owner or consultant may desire additional information or clarification from the contractor, but this should be limited to valid requests related to the project.

The owner normally reserves the right to reject all bids, but this typically happens only if all bids come in over the budget or there are substantive irregularities in several of the low bids. If, during the bidding process, serious errors in design are revealed, bidding should be stopped before completion and all bids returned. If only one bid is received, the owner should return the bid unopened or be prepared to accept the bid if it is within the budget and the bidder is qualified (Smith 1987). Negotiating with the low bidder for a reduced quote or revised project scope is a tricky issue and is done less often with public projects than private; other bidders might have legal grounds to protest. In short, negotiating should be avoided, but if circumstances warrant it, a construction contract attorney should be involved.

If the lowest bid is withdrawn or the bidder is found to be unqualified, the next lowest bid is evaluated and the bidder's qualifications reviewed. However, bids are valid for a specific time period, and if more time is needed, an extension must be approved.

Contract award. The successful bidder is officially notified with a notice of award, and the contract documents are issued for signing. The contractor usually has a set period of time to execute the documents. When contract documents are signed and then countersigned by the owner, the contractor is sent a set of the signed contracts and a notice to proceed. The notice usually establishes the effective date of the beginning of the contract time.

REFERENCES

American Society of Civil Engineers (ASCE). 1990. *Quality in the Constructed Project: A Guide for Owners, Designers, and Constructors*. Manuals and Reports on Engineering Practice No. 73. New York: ASCE.

Associated General Contractors of America (AGC). 1990. *Construction Contractor's Qualification Statement for Engineered Construction*. Document No. 220. Washington, D.C.: AGC.

Smith, R.J. 1987. *Recommended Competitive Bidding Procedures for Construction Projects*. Engineers Joint Contract Documents Committee (EJCDC) Document No. 1910-9-D. Washington, D.C.: American Consulting Engineers Council; New York: American Society of Civil Engineers; Alexandria, Va.: National Society of Professional Engineers.

Chapter 5

Performance Clauses

Completing a project on schedule is often as important as the cost and quality. For example, a new facility might be needed before seasonal peak demands, to serve a new industry, to meet the scheduling of subsequent projects, or to meet regulatory deadlines. The project schedule also affects the finances of a project because the interest "clock" on bonds is usually based on the proposed completion date. If the project is not completed on time, the interest cost accrues while revenues may be lost.

Because the completion schedule is so significant, the monetary worth of on-time completion should be estimated realistically and the effects of early and late completion evaluated. Where late completion will cause the owner to suffer damages, the contract terms should address how damages will be assessed. When early completion benefits the owner, the contractor should be appropriately rewarded. The traditional incentives used are liquidated damages and bonus and penalty clauses in the construction contracts. Other incentives, not used as often, are reduced retainage on progress estimates and delay compensation.

This chapter defines liquidated damages, bonus and penalty clauses, delay compensation, and reduced retainage. It also discusses other contractual elements affected by incentives, including progress payment estimates, quality of work, time definition and bid evaluation, and change orders. Examples of performance clause wording are included.

TYPES OF PERFORMANCE CLAUSES

Several types of contract clauses are used to give the contractor an incentive to complete the project on schedule or early. Such clauses are generally called performance clauses.

Liquidated Damages

The term liquidated damages refers to a specific sum of money paid to the owner by the contractor if the project is not completed on time. The amount is usually specified on an amount-per-day basis. Essentially, liquidated damages represent a forecast of the financial damage to the owner caused by failure to complete the contract on time. Liquidated damages are not the same as a penalty (as discussed later).

The use of liquidated damages is appropriate when actual damages are hard to calculate and prove. It is difficult to change the liquidated damage amount after the contract is signed; courts are reluctant to set aside the initial agreement. However, if the sum stated in the contract for liquidated damages is highly disproportionate to the damages actually incurred, courts may examine the circumstances and interpret the sum to be a "penalty" and declare it void, especially where the bargaining power between the parties was very unequal.

Liquidated damages clauses should never use the term *penalty* and must set forth sound reasons why the parties chose to use a liquidated damages clause. The courts have ruled to present a distinction between a penalty and liquidated damages. Two criteria seem paramount in determining whether the monetary amount is for the purpose of securing performance of the contract (and is therefore a penalty) or whether it is intended to be paid in lieu of performance (and therefore represents liquidated damages). The criteria, presented in the form of questions, are (1) at the time of the making of the contract, would the sum provided seem to bear any reasonable relationship to the anticipated damages? and (2) would the actual damages be difficult or impossible to ascertain? If both answers are yes, the sum provided would normally be considered liquidated damages.

Another important factor for liquidated damages is the ability to quantify the damages at the time the contract is signed. Whenever all or a portion of the delay is due to an unforeseeable cause beyond the control of the contractor, the courts will not enforce liquidated damages for that portion of the delay. Strikes, fires, delayed delivery of materials, actions of the owner, or natural disasters are the types of events that may excuse a delay. Often owners are faced with spending time and money on legal fees proving a damages claim. Some will just drop the claim and disqualify the contractor for future projects. Examples of liquidated damage clauses are included at the end of this chapter.

Bonus and Penalty Clauses

Bonus and penalty clauses are used solely as incentives to finish a project early or on time. A bonus clause is an amount, usually per day, to be paid by the owner to the contractor for work completed before the completion date. The penalty is an amount, usually per day, paid by the contractor to the owner when the project is not completed on time.

Unless actual damages occur, as in liquidated damages, many states will not enforce a penalty clause without a corresponding bonus clause. In many cases, courts will then limit the amount of liquidated damages to the actual damages incurred, up to the maximum amount stated in the contract. The presence of a bonus clause will not preserve a liquidated damages amount if the liquidated damages sums are unreasonably more than the actual damages. Examples of bonus-penalty clause wording are found at the end of this chapter.

Delay Compensation

Contractors are increasingly seeking delay compensation in contracts that contain liquidated damages or penalty statements. Delay compensation can be thought of as liquidated damages in reverse. The owner would pay the contractor an amount, usually per day, if the project is not completed on schedule because of delays caused by the owner. A cap can be established on delay compensation. Unless there is a cap in the contract, a contractor could demand delay damages many times greater than the actual "hard cost" damage. The owner's attorney should determine the legality of such a cap in each respective locale.

Reduced Retainage

Retainage refers to contractor earnings withheld in case (1) there was an error in estimating the level of work the project required, or (2) poor quality work or errors were revealed that could not be determined until subsequent work. The retainage is usually a percentage of the pay estimate, sometimes with a maximum amount specified. The total retainage is usually paid after completion. Chapter 14 discusses retainage in more detail.

Reduced retainage may be used as an incentive for early completion. The contract can be structured such that upon satisfactory completion of key items, or at a certain phase of the contract (such as 50 percent complete), the retainage percentage or maximum amount can be reduced. However, if the owner releases the retainage and there is subsequent damage or default, the owner usually can recover the retainage only if the amount of money that remains unpaid on the project is greater than the retainage itself.

OTHER PERFORMANCE INCENTIVES AND ISSUES

Other incentives for on-time or early completion are available to the owner. These incentives are not usually written into contracts.

Timely Progress Payments

The timeliness of progress payments to the contractor may serve as an incentive for contractors to maintain a contract schedule. If payments are made in a timely manner after a pay estimate is submitted by the contractor and accepted, for example, in 10 days instead of 30, the contractor usually will keep the project moving on schedule. Such timely payments help the contractor recover or reduce actual, administrative, and interest costs.

Time Definition

When bonus-penalty, liquidated damages, or delay compensation clauses are incorporated into contracts, calendar days or working days must be specified as the time basis. Calendar days count every day; working days are Monday through Friday, not including holidays. Calendar days are the easiest to use and the most common practice.

Bid Evaluation

With some projects, bidders must include a time of completion with their bids. If a contractor is the low bidder but has significantly more time for completion proposed in its bid than the others, is this the actual low bid? Obviously the answer depends on whether time is of the essence for the contract, a criterion that should be clearly established in the bid documents. Assuming time of completion is of the essence in a project, one way to evaluate bids is to compare the low bidder's time of completion against those of the next lower bidder's. Calculate a bonus to be paid to the low bidder based on the completion date of the next lower bidder. With the bonus added to the bid, determine if the bid is still the lowest. For public bidding the bonus amount and procedure should be determined and announced to bidders in bidding documents; it may be difficult to justify after the bids are opened. An owner may elect not to have a bonus or penalty in effect for a short period, for example, two weeks before or after the scheduled completion date.

Change Orders

Change orders can significantly delay a project's completion. When a change is proposed, the time necessary to complete the change has to be realistically determined. If the contractor is ahead of schedule and the proposed change adds time to its early completion schedule, arguably the original schedule should be extended so that the contractor maintains its original initiative for working toward the bonus. If the contractor is behind schedule and the proposed change would decrease working time, the owner should consider whether the original completion date needs to be revised to maintain consistency.

Quality of Work

To a lesser extent bonus and penalty clauses may be applied to the quality of work. Applying the clause to quality is difficult and may require the use of statistical methods of evaluation as well as a compromise between contract specifications and actual results. For example, if a steel structure receives a coating in excess of that specified, should the contractor receive a bonus for exceptional work? Alternatively, when the coating is less than that specified, should it be accepted and a penalty assessed, or should the contractor redo the work at its own expense according to the minimum specifications written in the contract documents? Each case is different and should be assessed on its own merits.

Similarly, should the contractor be rewarded for installing a more efficient system than specified without a negotiated increase in cost, or should such items remain in the realm of negotiated voluntary alternatives? A clause in the contract that the owner and contractor will share on cost savings for acceptable alternatives to the contract documents may be one approach to answering these questions.

EXAMPLES

The following are examples of standard liquidated damages and bonus and penalty clauses.

EJCDC Documents

Many water utilities, consultants, and contractors use EJCDC documents in design and construction projects. A history of the EJCDC and a list of relevant documents appear in chapter 2.

3.2 Liquidated Damages. OWNER and CONTRACTOR recognize that time is of the essence of this Agreement and that the OWNER will suffer financial loss if the Work is not completed within the times specified in paragraph 3.1 above, plus any extensions thereof allowed in accordance with Article 12 of the General Conditions. They also recognize the delays, expense, and difficulties involved in proving the actual loss suffered by the OWNER if the work is not completed on time. Accordingly, instead of requiring any such proof, OWNER and CONTRACTOR agree that as liquidated damages for delay (but not as a penalty) CONTRACTOR shall pay OWNER____dollars ($____) for each day that expires after the time specified in paragraph 3.1 for Substantial Completion until the Work is substantially complete. After Substantial Completion, if CONTRACTOR shall neglect, refuse or fail to complete the remaining work within the time specified in paragraph 3.1 for completion and readiness for final payment or any proper extension thereof granted by OWNER, CONTRACTOR shall pay OWNER___ dollars ($___) for each day that expires after the time specified in paragraph 3.1 for completion and readiness for final payment.

Indianapolis Water Company Standard Forms

The following two standard forms are specific to the Indianapolis Water Company.

SC-18 Liquidated Damages. It is understood and agreed that time is of the essence of the contract. Should the Contractor fail to perform the work within the period of time stipulated in the Contract Agreement, the Contractor shall pay to the Owner, as liquidated damages and not as a penalty, all inspection costs and other expenses accruing to the Owner, but not to exceed $500.00 per calendar day of default *unless* extensions of time granted by the Owner specifically provide for the waiving of liquidated damages. The Owner shall have the right to deduct the liquidated damages from any moneys in its hands, otherwise due, or to become due, to the Contractor, or to sue for and recover compensation for damages for nonperformance of this Contract within the time stipulated.

SC-19 Bonus Payments. It is understood and agreed that time is of the essence of the Contract. In the event that the Contractor substantially completes the work ahead of the Contractor's agreed substantial completion date as established in the Bid Form, the Contractor shall receive as additional compensation $500.00 per each calendar day ahead of the substantial completion date. In addition, on the pay estimate following the substantial completion date, the retained percentage on those items substantially complete shall be released.

Discussion

Bonus payments, at times called incentive payments, may have a maximum amount specified. In this case the owner can set a maximum value that early completion is worth. This limit is used more often when the construction schedule is not tight. It is important to note the wording in the liquidated damages examples; the damages must be clearly stated as damages and not as a penalty. In the EJCDC document the wording acknowledges that proof of damage is needed but that, because of the expense of proving it, both parties agree that a certain sum will be paid to avoid costly legal fees. In the Indianapolis Water Company example it is implied that the owner will bear the burden of proving the cost of the project not being done on time.

Nontraditional Construction Contract Administration

Owner Construction Management

Construction Management Firms

Multiple Prime Contracts

Design-Build and Turnkey Construction

Guaranteed Maximum Price Construction Management

Chapter 6

Owner Construction Management

Some degree of management is required of each of the participants involved in the design and construction of any engineered project—the owner, the design consultant, and the contractor. This chapter examines owner construction management, which refers to an owner that assumes all the responsibilities except for the final design and construction of the project.

CHARACTERISTICS AND ACTIVITIES

With owner construction management, the owner has made a decision to be the principal member of the project team. This form of management can vary from utility to utility, depending on staff capabilities. For this discussion the owner is assumed to be active in all phases of the project. Specific owner activities are discussed in the following paragraphs.

The owner guides the design consultant and makes certain that the completed project will serve the intended purpose. The owner is more involved in the final design and often performs many tasks such as soliciting bids on the work, awarding the contract, and interpreting all details to the contractor with a minimum of assistance from the design consultant. Owner construction management can facilitate the use of multiple prime contracts, as discussed in chapter 8.

After awarding contracts, the owner must assume additional responsibilities including

- construction inspection
- materials testing
- evaluating contract changes that may be required
- negotiation for and documentation of change orders
- reviewing contractor's submittals for labor expended and materials installed for progress payments

Figure 6-1 Steelworkers on 6-mil gal (22,710-m^3) reinforced concrete tank in San Diego County, Calif.

Throughout the construction period, safety on the jobsite must be a top priority. As contract administrator, the owner must make sure the contractor is following safety regulations, such as the scaffolding shown in Figure 6-1. Because of the owner's additional exposure under this form of management, the owner must be protected both contractually and by appropriate insurance coverage. An insurance consultant or a project engineer in the owner's management team could be a wise investment.

The owner and contractor establish a completion date and time schedule. The owner must perform its own responsibilities (such as testing and inspection) in a timely manner. As work is completed, the owner assembles accurate as-built drawings from contractors, as well as operations and maintenance (O&M) manuals from both contractors and equipment suppliers. These must be carefully filed and preserved for use in developing ongoing maintenance programs and general operating procedures required for the completed facility. The owner may also need to negotiate maintenance contracts for complex equipment or for units such as elevators for which the safety of the public is a special consideration.

Each construction management activity shares tasks in common: problem identification and alternative selection, cost estimation, project approval, preliminary design, final design, construction, closeout, and final asset entry. To control all of this activity, the owner must develop internal systems that permit the necessary communication among all groups, particularly construction management, engineering, finance, purchasing, and records. All of the systems must have access to information concerning the current status of the work during planning and throughout construction.

STAFF REQUIREMENTS AND RESPONSIBILITIES

Of primary concern when a utility decides to perform the construction management phase of a project is the availability and expertise of its staff. Even with relatively simple projects, construction management can take extensive staff time. Construction management itself is an art as well as a science, and a good practitioner is usually one who has much experience.

The owner usually performs an evaluation of its staff availability and expertise during the planning phase of the project. A typical evaluation involves three steps:

1. Estimate the number of hours and type of expertise involved for each phase of the project.
2. Compare the result to the existing staff's availability and expertise.
3. Determine whether existing staffing is adequate, new staff (temporary or permanent) should be hired, or another method of contract administration should be used.

Table 6-1 shows an example of following these steps. In this case, the owner has adequate staff to cover all aspects of the project except for some engineering personnel needed after the preliminary approval stage.

Unfortunately, no rules of thumb have evolved to help estimate the time involved in project management. The variables are too dependent on the project characteristics and the expertise of the staff. The best way to arrive at good estimates is to compare the current project to similar past projects. To help with the estimation, the following paragraphs describe the owner's staff responsibilities during the project phases.

Table 6-1 Example of estimating project staff requirements

Activity	Administration and Planning			Purchase and Finance			Engineering			Maintenance and Operations		
	Tot	Stf	Req	Tot	Stf	Req	Tot	Stf	Req	Tot	Stf	Req
Concept-project development	400	400	0	200	200	0	200	200	0	100	100	0
Preliminary design	100	100	0	80	80	0	400	400	0	40	40	0
Scheduling	40	40	0	80	80	0	80	80	0	0	0	0
Cost estimating	40	40	0	80	80	0	80	80	0	40	40	0
Preliminary approval	80	80	0	40	40	0	80	80	0	0	0	0
Final design	100	100	0	80	80	0	600	200	400	40	40	0
Approvals and permits	80	80	0	0	0	0	120	80	40	0	0	0
Solicitation of bids	0	0	0	40	40	0	80	40	40	0	0	0
Award of contract	20	20	0	20	20	0	40	20	20	0	0	0
Construction activity	60	60	0	60	60	0	800	300	500	100	100	0
Closeout	40	40	0	40	40	0	200	100	100	40	40	0

Tot - total time required per activity in hours.
Stf - staff time available in hours.
Req - required additional time (design engineer and/or new employees) in hours.

Planning

Planning begins with the formation of the owner's management team. Most often a member of the owner's engineering or administrative division will head the management team. Team members can come from all the utility's divisions, including operations, engineering, administration and human resources, and finance.

The team completes the preliminary study, which defines the need for the project, determines possible alternatives, estimates costs and personnel requirements, and determines financial capability. Results of the preliminary study should help the utility decide if the owner can manage the project. The management team then completes the final report, which refines the project scope and budget and serves as the basis of design.

Design

The project management team is usually charged with selecting the design consultant. Previous chapters have discussed the selection procedures. Once selected, the management team must establish a procedure for working with the consultant, usually by establishing a particular staff member to serve as contact for the consultant.

The staff contact person is usually empowered to make decisions for the utility as far as the design is concerned, unless the question is beyond that individual's power or ability to answer. If such a question arises, the contact member finds the staff member who can provide the answer. The contact member is also responsible for setting up the necessary meetings between the consultant and the management team or with any other staff member. The staff contact usually facilitates in-house review and approval of plans and specifications.

During the final design phase, the owner may wish to contract for land-clearing, geotechnical, and similar services required by the designer and to permit an early start of actual construction. Utility permits can be obtained during this period so that buried pipe installations can be completed early if this is in keeping with the overall program and schedule.

Usually the contract and bidding documents are prepared by the consultant; however, the owner may have forms it prefers to use, and using them may speed up the process because the owner will not need as much time for review. Likewise, the actual bidding may also be conducted by the owner or the consultant, depending on the owner's preference. Previous chapters discuss bidding procedures in detail.

Construction

With owner construction management, the owner has several responsibilities during the construction phase. The owner must provide staff who have the time and qualifications to expedite, not hinder, construction. The main categories of responsibilities of the owner project manager are as follows:

- Perform required inspections.
- Perform (or subcontract) materials tests.
- Process progress payment requests.
- Provide contractor with plan and contract clarification.
- Review shop drawings along with the designer.
- Conduct and attend preconstruction and construction progress meetings.
- Make sure that the owner's employees perform their tasks safely when on-site. Any possible unsafe working methods on the part of the contractor's

crews or subcontractors should be reported to the contractor. (NOTE: Exactly how much the owner is responsible and liable for jobsite safety should be reviewed by legal and insurance experts.)
- Review contract changes and process change orders.
- Prepare (or review) record drawings (also called "as-builts").
- Obtain or develop O&M manuals and procedures.
- Furnish certification of completeness to the contractor and applicable regulatory agencies.

With owner construction management, special training for the owner's staff may be required, in particular the construction observers or inspectors. Construction inspectors must be aware that their primary responsibility is to see that the project is completed in compliance with the requirements of all other contract documents. In other words, the owner's construction inspectors should not be changing the design in the field. Following clearly established procedures for change approval is important.

Construction inspectors should also be trained in keeping a detailed daily diary of progress. Such diaries are often crucial in settling disputes, as well as in solving field problems. Photographs or video should be taken. Good construction photographs show the work in various stages and at different angles such as the series of photographs of a water tower installation in Figure 6-2. Useful photographs should also show the date and scale of objects in the picture, such as in Figure 6-3, where two workers in the picture provide scale to the steel reinforcing.

Items to be noted in daily progress diaries include
- construction activities
- site conditions and weather
- what materials tests were made
- what inspections were made
- what problems were encountered
- who was informed of the problem
- how the problem was resolved

With multiple prime contracts (chapter 8) or special equipment contracts, the owner must coordinate delivery and installation with the other contractors. Coordination may be assigned to one of the contractors through construction documents. These types of contracts are typically used by more experienced owners on larger projects. Multiple contractors can speed up work and give the owner more control over project quality, but they also require more knowledge and ability.

As contracts and procurement agreements are completed, work must be inspected for final acceptance and contracts must be closed out. A punch list of construction items that need correction or changing is used to indicate items that may not be complete or require corrective work. With owner construction management, the owner's direct involvement in the work should reduce the number of punch list items.

At this point the owner construction manager must complete as-built drawings or receive them from the contractors and verify their accuracy. These drawings, together with O&M manuals on equipment units, must be filed and shared with operating divisions for use in developing O&M programs.

Because each project should be viewed as a learning experience, project documentation should be stored. Numerical data can be entered in an electronic database for easy access and comparison. Data and other documentation can be used to plan, estimate costs, schedule, and manage future projects. Documentation should include
- project description
- planned and actual completion dates
- unit costs for the work

Figure 6-2 Stages in the construction of an elevated steel tank needed to maintain the necessary head for filter backwash water

- cash flows broken down into materials, labor, contractor, and miscellaneous costs
- owner's labor and materials costs
- photo records and diaries

ADVANTAGES

Owner contract administration has several advantages, particularly if the owner has trained staff with the necessary time to devote to the project. The following paragraphs present important advantages and what the owner must do to maintain them.

Figure 6-3 Expansion of conventional water treatment plant from 3 to 6 mgd (0.1314 to 0.2629 m^3/s) in Auburn, Calif.

Direct Communication

Rather than spreading out the responsibilities among several participants, especially to an outside contract administrator, the owner keeps them in-house. Communication between owner and contractor is more direct, without another participant acting as a go-between. The owner's staff is more familiar with all the utility's systems and can reduce the potential for problems.

For this advantage to work, the owner must maintain open and direct lines of communication within its own organization as well as between its staff and those of the principal designer and contractor. Direct communication (see chapter 12) facilitates final design, construction, early completion, acceptance, and startup of the new facility or installation.

Smoother Process for Final Design and Approval

When the owner is more involved in the project from the start, particularly if its staff performs preliminary design, the owner has developed a stronger conceptual image of the project. With a strong concept, the owner can better communicate its goals and requirements for the project, which helps both the final design and construction.

When the owner is actively involved in the design process, the owner can obtain permits to arrange subsurface explorations and analysis, site clearing and similar activities, and possibly excavation. The need for rights-of-way can also be determined, and legal and contractual activity can be expedited.

Timely processing of design, changes, and shop drawing approvals is essential. It is important throughout the construction phase of the work, but it is especially important at the beginning, because of the greater volume of information that must be approved to permit work on the contract to begin. The active owner is in a better position to complete these reviews.

Improved Resolution of Construction Problems

Many problems that arise during construction are a result of differences in interpreting plans. The owner who has been active in the preconstruction and construction phases can more readily resolve such differences. If changes to the design are needed, the owner can work directly with the design consultant, rather than working through the construction manager. If a cost increase is involved, the active owner can more readily decide how to pay for it.

Involvement of Operating Personnel

The active owner's personnel who will operate and maintain the facility can more easily be involved from the beginning. This involvement can have a number of advantages, including the following:

- Early development of the team concept by the owner's personnel can increase open communication.
- The availability of persons knowledgeable in the owner's requirements, operations, and fiscal limitations can ensure the intended detail in design, proper interpretation of the contract documents in plan review and approval, and timely solutions for day-to-day problems that arise during construction.
- Operations staff who have been involved in the work are in a better position to bring the new facility on-line in the shortest possible time, to keep unscheduled downtime in this critical period to an absolute minimum, and to assist in training required additional workers.

DISADVANTAGES

The main disadvantage of owner construction management is the staff time required. Most utilities cannot take staff away from their normal day-to-day tasks. Similarly, project management requires special skills and experience that existing staff may not possess.

The owner assumes more of the risk of the project when it serves as project manager. Furthermore, an outside project manager may have a more objective view of the project or may bring broader experiences that might allow more creative and efficient problem solving.

Chapter 7 discusses an option for an owner that does not have the staff with adequate time, expertise, or experience to perform contract administration; that is, the use of a construction management firm.

Chapter 7

Construction Management Firms

If the project owner's staff does not have enough time, expertise, or experience, the owner can hire a construction management firm (CM) to perform contract administration. This chapter presents characteristics of a project managed by a CM firm, advantages and disadvantages of using such a firm, project participants' responsibilities, and contract administration procedures.

CHARACTERISTICS

A project owner can contract a construction management firm for its special expertise in the construction phase of a project, particularly with complex projects such as those shown in Figures 7-1 and 7-2. For these cases, a CM usually does not provide design or construction services; the owner contracts directly with the design consultant and the construction contractor. The CM firm does maintain a peer relationship with the design consultant and the construction contractor performing the work. There are also administrative ties among the design consultant, the construction contractor, and the construction manager.

The CM's main responsibility is to perform contract administration during the construction phase of a project. In a traditional project (chapter 1), the owner or design consultant performs such construction management tasks. In this chapter, the CM is assumed to be a separate entity, that is, not the design consultant.

The CM's responsibilities (expanded on later in this chapter) include observing construction, scheduling inspections, resolving field problems, and processing change orders and progress payments. The CM may also have responsibilities in the planning phase, the design and bidding phase, and the postconstruction phase. However, these responsibilities are not typical of CM-managed projects.

Figure 7-1 Pump station (100 mgd [4,381 m^3/s]) in Orange County, Calif.

Figure 7-2 Fort Myers, Fla., 12-mgd (0.5258-m^3/s) nanofiltration water treatment plant

Payment to the CM is usually on a fixed fee or cost-plus basis. A cost-plus-fee approach can include specification of a maximum total payment. Any payment greater than the maximum would have to be justified, such as in the case of an expanded project scope or increased work due to unforeseen site conditions.

ADVANTAGES AND DISADVANTAGES

Proper construction management requires a commitment of time and a high level of expertise and experience. Appropriate construction management can make the difference between a successful project and one less than successful.

The owner of a project must determine whether its staff has the time, expertise, and experience to provide proper construction management. In chapter 6, a simple method of making such a determination was discussed. If the owner determines that its staff does not have the time, expertise, or experience to perform construction management, the owner can hire the additional staff required or retain the services of a construction management agency.

Either of those two options requires funding, and the owner must look at the economics of the options. For example, if the owner hires new staff, the cost of benefits and other overhead must be included in their cost to the project. And what happens to the staff after the project is completed? Will they become permanent employees? If so, permanent positions must be justified. If additional staff is hired on a temporary basis, it may be difficult to find qualified people.

Advantages

The main advantages of hiring a CM are

- cost savings
- quality control
- risk management

Cost savings. Retaining the services of a CM may be less expensive to the owner than using the staff time required for contract administration or hiring additional staff. The cost of the CM may also be considered part of the capitalized cost of a project and could be covered under the project financing such as bonds. Staff time may or may not be so covered.

A construction management firm, through its expertise and experience with the type of project to be constructed, should be able to help control project costs. For example, during construction, the CM can quickly resolve field problems and endeavor to keep changes to a minimum. If the CM is retained early in the project, it could add its expertise in the planning and design phases, where most of the decisions affecting cost are made. In particular, the CM can help with material selection, contract packaging and selection, project staging, and sequencing.

Quality control. During construction, the CM may be more objective than the owner. For example, an outside firm might possibly be more demanding on a local contractor than an owner who often works with that contractor. The agency may have had a wider range of experiences and can better anticipate and deal with potential problems than can the owner.

Risk management. By hiring a CM, the owner transfers some of the project risk to that firm. The CM can be held responsible for the project quality and cost consequences of its work as contract administrator.

Disadvantages

The CM does add another participant to the construction process, which may require additional administration and coordination by the owner. A poorly performing CM creates many problems for the owner, in which case owner contract administration would be preferable.

RESPONSIBILITIES AND PROCEDURES

The following discussion assumes that the CM has been hired for contract administration during the construction phase, as is usually the case. In rare cases the CM may be hired earlier in the project. For a discussion of CM responsibilities in other phases of a project, refer to chapter 10.

Owner

The owner must clearly establish the responsibilities of the design consultant, construction management firm, and contractor. Without a clear understanding, tasks will go undone and conflicts will inevitably occur.

The owner, having decided to hire a CM, goes through the same procedure as with selecting a design consultant. Those procedures were discussed in chapter 1. The important thing for the owner to remember is to hire the CM well before the construction is scheduled to begin. The CM must have enough time to become familiar with the project, as well as to organize its staff. On complex projects, this lead time could be anywhere from several weeks to several months.

During the construction phase, the owner depends on the CM for day-to-day contract administration of the project, especially coordination and communication. The owner must provide the staff necessary to perform several duties, such as

- resolving field problems
- reviewing and approving changes
- processing pay requests
- attending construction progress meetings
- observing startup and testing
- providing staff for training

Construction Management Firm

The construction management firm's specific responsibilities vary considerably from project to project. As always, these responsibilities must be clearly established and agreed to by all participants. For example, Figures 7-3A, 7-3B, and 7-3C show the installation of pipes in a chlorine contact structure, a task that requires careful monitoring.

The important responsibilities that must be established include

- field authority or line of authority for making decisions
- inspections and materials testing
- construction monitoring
- environmental and safety protection
- community relations

Figure 7-3A Installation of 54-in. (1,350-mm) discharge line in chlorine contact structure

Figure 7-3B Installation of lines in chlorine contact structure

Figure 7-3c New discharge pipes in chlorine contact structure

After being hired by the owner, the CM must become thoroughly familiar with the project, including

- drawings and specifications
- contract conditions
- site conditions
- owner's current systems, especially at connection points with the project

The CM usually has a site representative (also called the resident manager, or project manager) who is always at the project and who is responsible for contract administration, coordination, and communication. These responsibilities are largely the same as discussed in chapter 6 for owner contract administration; however, there are some differences as can be seen when the following list of CM responsibilities is compared with the list in chapter 6:

- Perform required observations and inspections.
- Coordinate inspections to be performed by owner's staff.
- Perform (or subcontract) materials tests.
- Process progress payment requests from contractor (to be submitted to owner).
- Provide contractor with plan and contract clarification, if within the CM's abilities; otherwise refer the question to the owner or design consultant.
- Conduct and attend construction progress meetings.
- Make sure that the CM's employees perform their tasks safely when on-site. Report any possible unsafe working methods on the part of the contractor's crews or subcontractors to the contractor. (NOTE: Exactly how much the CM is responsible and liable for jobsite safety should be reviewed by legal and insurance experts.)
- Review contract changes and process change orders to be approved by owner.
- Prepare (or review contractor's) record drawings.
- Furnish certification of completeness to the contractor and applicable regulatory agencies.

The CM's primary responsibility is to see that the project is completed as shown on the project drawings and in compliance with the requirements of all other contract documents. All design or other changes must be approved by the owner and design consultant.

As discussed in chapter 6, construction inspectors and observers should also be trained in keeping a detailed daily diary of the progress. Refer to chapter 6 for a list of the events that should be recorded. The CM usually keeps diaries and other records, rather than giving them to the owner. In the event of a dispute, the agency may be required to give a copy to the owner and other participants.

The CM normally works with the owner to develop the punch list of final completion items. The agency usually prepares the as-built drawings to be given to the owner.

Design Consultant

The design consultant usually does not perform construction management but does perform the other typical tasks required of the designer, including

- shop drawing review
- design change review and approval
- field problem resolution

The design consultant rarely has a site representative on a CM-managed project; however, the designer should have a staff contact available at all times during construction.

Chapter 8

Multiple Prime Contracts

Multiple prime contracts (MPC) construction refers to a project that has more than one construction firm contracting directly with the owner. The two principal categories of MPC are the state-mandated MPC and the construction management approach. This chapter discusses each of these categories and looks at the advantages and disadvantages of multiple prime contractors versus a single prime contractor.

MULTIPLE PRIME CONTRACTS PROJECT MANAGEMENT

As with all projects, the owner must select an effective contracting approach to complete the project on time and on budget and to meet specifications. Particularly with complex projects such as a treatment facility, one approach is to use multiple prime contractors. With MPC, the owner contracts specific portions of the project to specialty contractors, most often in the following categories or trades:

- general construction
- electrical
- heating, ventilation, air conditioning (HVAC or mechanical)
- plumbing

In some US states, MPC is required; such a situation is referred to as state-mandated MPC construction.

The MPC approach is better suited to medium- to large-scale projects, mainly because of the increased management costs for the owner. MPC is very attractive when the bidders for the project do not have the size, resources, and track record to prove their ability to handle the entire project successfully and independently. If such a case occurs, splitting the project into smaller pieces can be advantageous to the owner.

State-Mandated MPC

Separate prime trade contractors are required in New York and North Carolina, among other US states. In North Carolina, for example, state-mandated prime trade construction is required if the value of the work involved in a particular trade is in excess of $30,000. Some states also allow single prime bids, in which case the owner

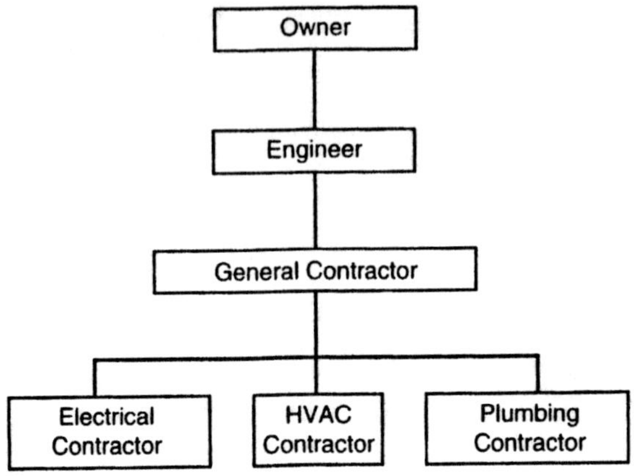

Figure 8-1 Typical flow of information for multiple prime contractors

would be allowed to accept either the low single prime bid or the low multiple prime group bid, whichever presents the lowest overall cost to the owner.

As mentioned earlier, the normal contract categories are general, electrical, HVAC, and plumbing. Even though the prime trade contractors directly contract with the owner, the general contractor usually coordinates all construction activities on a project. The general contractor needs to work closely with the prime contractors to ensure that all work is scheduled properly and installed and completed as specified, within cost and schedule constraints.

In addition, the owner typically hires a consultant or engineer to provide design and management services. In this case, the proper flow of information (though not necessarily authority) for an MPC project would be as shown in Figure 8-1. Figure 8-2 illustrates the flow of information if the project also has a construction manager.

The state-mandated MPC approach complies with laws and regulations that govern municipal procurement and awarding of contracts. All of the construction, materials, and equipment are awarded on the basis of competitive, fixed-price bids. Other types of contracts discussed in this manual, such as design–build or guaranteed maximum price, may not comply with such laws. An example of using MPC to satisfy such laws follows:

The project was to build a water treatment plant, with a 6-mgd capacity producing clarified water (0.5 ntu) from a high-turbidity river water supply. The project for the San Patricio Water District in Ingleside, Texas, included all treatment plant process components, buildings, aboveground and belowground piping, process control and instrumentation, pumping, electrical, water storage tanks, sludge-handling facilities, and all site improvements.

The time available from the beginning of final design to completion and operation of the plant was 34 weeks. A consulting engineer completed the preliminary design work. The engineer then joined the owner's staff, acting as owner's representative and coordinating the work of the multiple prime contractors. The MPC approach was selected because of the time limitations and the need for targeted process equipment selection. Equipment selection was critical because it controlled

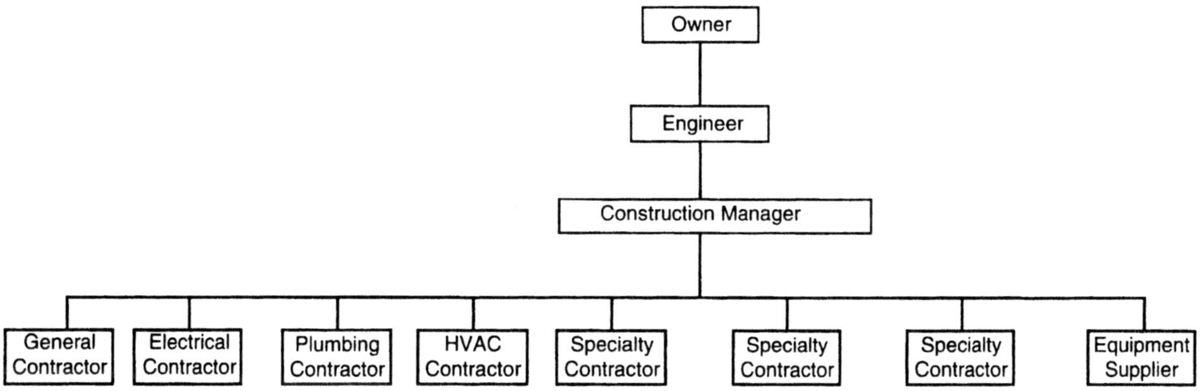

Figure 8-2 Typical construction management flow of information

building dimensions, underground piping sizing and location, electrical requirements, and other factors.

Separate contracts were issued for

- flocculation and settling process equipment
- filtration equipment
- pumps and motors
- underground piping, site work, and building and process unit foundations
- building, electrical, piping and process unit installation, and storage tanks

The owner and engineer were concerned about potential disagreements between contractors as to responsibilities. This concern did not turn out to be a problem, though, because the owner was involved on a direct and continuing basis. The advantage of including equipment manufacturers and construction personnel early in the process resulted in time and cost savings.

Construction Management MPC

The difference between construction management MPC and state-mandated MPC is that in the former case the owner has more flexibility on how to organize a project. In a project using the construction management approach, the project construction manager (CM) works with the owner and the design consultant or consultants to divide the project into a number of logical procurement and construction contracts for separate bidding. The CM may also be hired earlier in the project and may assist the owner in selecting the design consultant. In either case, the CM more closely manages the work of the trade contractors, particularly with respect to scheduling.

The following examples illustrate the CM-MPC approach.

Example 1 Oregon. A firm provided CM services on a \$106 million plant project in early 1978. Originally scheduled as a 5-year project, the time frame was extended to 7 years to accommodate the owner's ability to provide funds. Seven separate firms provided design services. The project was subdivided into 33 construction contracts, the largest of which was worth \$5.4 million including change orders, and 32 contracts for procurement of equipment and materials.

Only one significant claim was filed; it was settled for $37,000. Despite the funding-imposed delay in project execution, the project was completed in early 1985 at a total cost less than the original budget estimate. Demonstrated savings equaled the fee for CM services.

Example 2 Utah. A firm provided planning, design, and construction management services for a $110 million wastewater treatment facilities construction project. The project was divided into 17 construction and 8 equipment procurement contracts.

A benefit of subdividing this large project into smaller construction and equipment procurement contracts was a greater flexibility in scheduling. Working with a number of contractors made it easier to meet the owner's changing needs. Rather than a single firm having to make major scheduling revisions as timing changes became necessary, contractors were able to adjust their individual schedules more easily. This approach enabled the CM to accommodate a number of funding delays, as well as one funding acceleration, without disrupting the overall program.

Example 3 Tennessee. A firm provided scheduling, contract administration, and inspection services for $94 million worth of facilities construction. The program was divided into 91 construction contracts. At the midpoint of the program, construction bids on 38 contracts, representing $30 million in construction value, were $10 million below the budget estimate, allowing the addition of 12 construction packages to the total program.

CHARACTERISTICS OF MPC PROJECTS

The following paragraphs examine the characteristics, advantages, and disadvantages of MPC projects compared to projects using a single general contractor. The characteristics are typical but may not apply to every project.

Total Contract Price

The total contract price is often lower with the MPC process. Single prime contractors incorporate the subcontractors' bids and apply additional markups for bonds, insurance, overhead, and profits. This increased cost could vary significantly from project to project; but generally the increased markups can be as much as 5 percent of the total contract price, which can be significant on large projects. Another possible reason for the lower price is the increased involvement of the design consultants and the owner with the contractors to improve communication and reduce misunderstandings. The MPC process allows earlier involvement of specialty contractors and equipment and material suppliers, which also tends to keep prices lower.

Separate and Smaller Contracts

Smaller contracts allow more contractors an opportunity to bid on the job. The result is a more competitive bidding climate and often a chance for more local contractors to become involved in the project. Not only can this reduce project costs, but it may enhance community relations.

Smaller contracts usually allow contractors to estimate project costs more precisely. The need for a large, general contingency fund is reduced because individual contractors each have a clearly defined contingency. Contractors have more control, which may reduce the overall expense of contingencies.

Figure 8-3 Foundation work on dechlorination building

The use of MPC is inherent to the fast track process, in which construction can begin before design is complete. Typically, construction contracts for preliminary site work, foundation, and other underground work (Figure 8-3), plus procurement contracts for items of equipment requiring a long lead time, are awarded early in the process, while design continues on superstructure, interior finishes (Figure 8-4), final landscaping, and other parts. With proper scheduling and management, total project time and cost can be significantly reduced through the use of a multiple-contract, fast track approach.

Bid Shopping

Bid shopping is a process where the general contractor solicits lower quotations from the various subcontractors after its own bid is submitted to the project owner. The multiple prime contracting method prohibits postbid shopping by general contractors. Bid shopping can result in an "auction," which could drive the price of the subcontractors' work down to a point where they cannot financially perform. Although the project owner faces no financial repercussions, the project sometimes suffers from reduced quality. Some states require that subcontractors' bids be listed if they are greater than a certain amount.

Prepurchased Equipment

The MPC approach can require significant up-front planning. In the early phases of the project, the project team can prioritize project needs, for example, equipment items that have a long lead time. Such planning can reduce the duration of construction contracts—and the cost of contractor overhead and markup involved in each—resulting in an earlier completion of the entire project.

Figure 8-4 Finish work on expansion of conventional water treatment plant from 6 to 12 mgd (0.2629 to 0.5258 m^3/s)

Early purchase of equipment can also assist the designers in preparing more complete plans because specific equipment dimensions, performance, electrical requirements, and other parameters are known. This helps the owner plan for operations and maintenance of the equipment (staffing, parts inventory, lubricants). It makes it easier for the owner to obtain assistance from the equipment supplier's staff; suppliers are more eager to contribute their energies, ideas, and time to a contractor who has made a commitment to purchase equipment.

With the CM and MPC approaches, the early planning and subdividing of the project into various contracts promotes improved long-range planning.

Design Costs

Design costs tend to be lower with the single prime contractor method approach. Multiple prime contracts require the design engineer to define distinct separations of work to form a contractual basis with the multiple prime contractors. Additionally, the design engineer must create a separate bid and proposal form for each prime contractor.

However, it is important to note that MPC encourages increased involvement of the design consultants and owner. Furthermore, MPC promotes earlier involvement by specialty contractors and equipment and material suppliers, which may have a large impact on lowering costs.

The MPC approach often requires the design engineer to focus more on the details of the project in order to deal with each prime contractor. Such focus can result in plans that are more complete and have fewer errors. Any additional design costs are not usually significant relative to the overall cost of the project. However, additional costs might be more than compensated through reduced overall costs and by the value offered in the early involvement of contractors.

Contract Administration

Contract administration costs can increase if the multiple prime contracting method is used. Under the MPC approach, the owner and engineer must work with each contractor separately with regard to progress meetings, shop-drawing review, pay requests, change orders, claims and disputes, and general coordination.

Claims and Disputes

There tend to be fewer claims and disputes from a project under single prime contracting. The single prime contracting method allows the owner to deal with a maximum of two other parties in the face of a contractual dispute—the engineer and the single prime contractor. Multiple prime contracts become very complicated when one contractor delays another or in the case of any of the multiple prime contractors becoming insolvent.

If multiple contractors are used, it is less likely that large claims on a project will be generated. Although one or several contractors might need to make a claim, each claim may be smaller than the claim made by a single prime contractor because each of the smaller contractors will have a smaller piece of the project.

Claim settlement or negotiation (chapter 13) is based on communication between the parties and their mutual desire to reach agreement on the claim. It is easier to reach this mutual agreement when small claims, rather than large claims, are presented for negotiation. Effective, creative negotiation is greatly preferable to litigation.

Quality Control

Quality control can be easier with the single prime method for several reasons:

- The general contractor becomes another responsible party in the shop-drawing review and coordination process.
- The general contractor is responsible for the quality control of all the finished work product.
- The owner has one party to hold responsible for quality control.
- In the event of damaged work, it is not the owner's responsibility to prove which contractor damaged the work.

No price tag can be placed on quality. Needless to say, whichever method is chosen must create a project of high quality in order to meet the owner's needs. Although higher quality may be easier to achieve under the single prime approach, it can still be achieved by using multiple contractors, though additional coordination and effort are required by the owner and engineer.

The concept of total quality management (TQM) can be used in construction to allow every supervisor and worker the opportunity to contribute more actively to the successful completion of the work. TQM stresses the importance of customer–supplier relationships. The CM approach is also based on effectively managing these relationships. TQM emphasizes the efforts of the team, rather than of the individual, as does the CM approach. And, perhaps most important, both systems underscore the need for open communications and continuous improvement.

Communication and Coordination

Communication between the general contractor and respective subcontractors is often easier with the single prime contracting method. However, the direct line of communication between the owner, the engineer, and separate primes is improved in the multiple prime process. More direct and focused coordination meetings are necessary under the multiple prime process.

Most general contractors are accustomed to coordinating all work activities at the jobsite. General contractors are experienced in performing their own work activities as well as in coordinating a large number of suppliers and subcontractors. Therefore, overall project coordination tends to be better with the single prime contracting method.

Cash Flow Management

Multiple prime contracts can reduce cash flow constraints by allowing the owner to match the work placed under contract to the availability of funds. Given the high costs of borrowing money, this advantage can also reduce interest costs.

GENERAL QUALIFICATIONS AND PROCEDURES

For an MPC project to be successful, the owner, engineer, construction manager, and contractors should meet certain qualifications and follow certain procedures, as discussed in the following paragraphs.

Adequate Experience

The owner, engineer, and construction manager should be experienced in working under an MPC approach and must be adequately staffed. The owner, engineer, and CM in the multiple contract approach are taking on much of the scheduling, coordination, and control responsibilities that would normally be performed by a single prime contractor. These tasks require a great deal of experience to be performed well. Failure to perform can lead to increased claims for interference and delay, extensions of schedules, and an increased number of incidents between contractors where liability and responsibility lines become blurred. For projects with a cramped or tight project site or schedule, scheduling problems tend to increase.

Increased management costs may have to be justified to a utility's governing board. Costs should be shown to reduce overall contract price, reduce change orders and disputes, and reduce overall construction time.

MPC projects are particularly challenging for a general contractor (as opposed to a single prime contractor) because the general contractor has to determine a realistic and accurate schedule for the overall project. Obtaining scheduling information from other prime contractors may be difficult. The owner should schedule meetings to be attended by all prime contractors, to be certain that each contractor realizes the need for full cooperation and to provide schedule information essential to the overall planning and execution of the work.

In all projects work must be scheduled logically to avoid having multiple contractors in the same work space, or having to do work over. The electrical contractor, for example, might want to start installing duct banks as early as possible. These duct banks typically run all over the project site. If, however, the project is an entirely new one, it will be necessary for the electrical contractor to wait until the appropriate time to install duct banks. In this case, a number of preceding tasks must be completed before duct work begins. Site grading, sewer and drain installation, structural excavation, and structural backfill are some of these tasks. The owner or CM must be careful in working with contractors to ensure their schedules meet the needs of the overall project team. Even in expansion projects, it is crucial to sequence contractor schedules carefully.

REFERENCE

Mason, C.C. 1994. *Creating a Quality Company–Applying the Deming Philosophy*. New York: Random House.

Chapter 9

Design-Build and Turnkey Construction

Design–build is a construction method that uses a single entity for both the design and construction of a proposed facility. *Turnkey* is a term often used interchangeably with design–build; a turnkey project, however, implies that a single entity completes all the steps of a project, including such additional services as site acquisition, financing, startup, operations and maintenance (O&M), and staff training.

DESIGN-BUILD

The owner typically considers the design–build approach in response to one of the following conditions:

- an in-house lack of the expertise needed for successful project completion
- regulatory pressure to conform to a specified in-service date
- convenience of proceeding with construction before completion and approval of final construction plans

The design–build entity can be any of the following (Seigel 1995):

- a company that can perform design and construction in-house
- a joint venture between an engineer performing design services and a contractor performing construction
- a general contractor or specialty contractor who subcontracts design to an engineer
- an engineer who subcontracts construction to a contractor
- a developer or entrepreneur who subcontracts both design and construction

A design–build project is approached differently depending on whether the owner is a public or private entity. State competitive bid laws, local governmental approval processes, and required technical expertise usually dictate whether the design–build concept is an acceptable alternative for a construction project.

Public entities usually consider the design–build project approach when the required specialized expertise is beyond their staff capability or that of their consultant. The owner must define the scope of work to be performed by the

Figure 9-1 Elevated steel reservoir (0.75 mil gal [2,839 m^3]), city of Richardson, Texas

design–build project team, the acceptable specifications used for completion of required work, and the basis of payment for construction services and materials.

A public utility is charged with the task of protecting the public trust and is held accountable for costs associated with making improvements. This charge makes it important that the design–build team players understand their respective roles and execute them to the fullest extent possible. Typical public sector projects that fall into this specialized category are elevated tanks (Figure 9-1), ground storage reservoirs (Figure 9-2), surface storage reservoirs (Figure 9-3), sludge-holding tanks, and specialty equipment, e.g., ozonators or large pumps.

A privately owned utility is usually not as constrained by bid laws and cost constraints as a public utility, but it still must meet regulations and perform economically. Often, the privately owned utility must meet tight project schedules and cannot take up staff time for project management. The design–build project approach can reduce construction time and project costs.

Team Players and Responsibilities

The design–build project team consists of several entities that work collectively to ensure the success of the project. A typical team includes the owner, resident utility plant engineer (construction manager), and design–build contractor. The owner establishes the criteria for the work to be performed, including

- project scope
- preliminary schedule
- selection process for required design–build services
- construction administration responsibilities
- method of payment for services

Figure 9-2 Two 16-mil gal (60,570-m^3) buried prestressed concrete reservoirs in Orange County, Calif.

The resident utility plant engineer, typically an employee of the owner, monitors the design–build project to determine whether the scope and schedule are being met. Where in-house expertise does not exist, the owner may employ the services of a construction manager who has the required expertise. Usually it is beneficial to have this person involved as early as possible in the decision-making process of the project. In a design–build project, the resident engineer advises the owner about project criteria, including

- current available technology
- constructibility of designs
- cost estimating
- critical path scheduling
- contract documents
- permit acquisition
- evaluation of submittals made by the contractor for materials or equipment
- status of construction
- project coordination with other entities

The design–build contractor provides the owner with the required expertise by combining design and construction services. The contractor is responsible for preliminary and final design, obtaining the necessary approvals, performing all construction or hiring subcontractors, and supplying all equipment and materials.

Contracting for Design-Build Services

Several methods of contracting for design–build services have been used and are described in this section. The method selected usually depends on the extent of regulatory constraints placed upon the owner and the nature of the project.

Competitive bid contracts. Competitive bidding allows the owner to secure design–build expertise that can meet the project requirements at the least cost. Some state laws prohibit bidding for consultant services but require bidding for construction

Figure 9-3 Roller-compacted concrete dam (139-ft [42.4-m] high) in Ruidoso, N.M.

services. Public utilities can overcome this dilemma by including performance specifications in the contract bid documents. Performance specifications allow the design–build contractor to include the design as part of the construction contract.

Negotiated fixed-price contracts. The negotiated fixed-price contract is typically used by the private sector owner to select the design–build entity to perform the work based on demonstrated experience, technical ability, and cost. Through negotiation, the owner can evaluate the firm's past performance and staff design and construction expertise, as well as establish a schedule that is mutually acceptable with the design–build entity. Usually, the owner pays a preliminary design fee for project conceptual plans.

Upon acceptance of the concept plans by the owner, a contract fee is negotiated. The design–build contractor assumes total project control to alter the design after the contract is signed. Problems can occur when cost liabilities for scope changes or additions are defined, since the contract was negotiated before the plans were fully developed.

Cost-plus-fee contracts. Cost-plus-fee contracts allow the owner to pay the actual cost for the design–build entity to design, perform contract administration, and construct the proposed facility. The owner and design–build contractor negotiate a fixed fee added to this cost. The fixed fee can be a lump sum amount, percentage of cost, or guaranteed maximum price (see chapter 10). Cost-plus contracts require the owner to manage the projects closely so costs are justified and do not exceed the budget.

Effective Design-Build Procedures

No established method will address every aspect of a design–build project. Each project includes some unique aspects that must be addressed to ensure successful project completion. However, some guidelines help the owner define the work to be performed and provide for control during the project implementation. These guidelines are outlined in the following paragraphs.

Establishment of performance parameters. The owner should outline necessary performance requirements before hiring the design–build contractor. These requirements may include such things as flow capacity, acceptable material type or strength, facility storage, life factor, maintainability, and regulatory constraints. The owner should take these requirements into account for contracting or as a basis of negotiation with the contractor.

Usually, a public sector owner includes performance specifications as part of the contract bid documents. In this way, the design–build contractor assumes the responsibility of providing necessary design expertise for the project. Once the design–build entity is selected, final design plans can be produced. These plans should be signed, sealed, and submitted to the owner for approval before bidding or proceeding with construction.

Construction phase coordination. As part of the contracted performance criteria, the owner should require the design–build contractor to provide shop drawings before construction or installation. Shop drawings are signed and sealed by a design consultant. The contractor provides a sequence of construction before beginning construction.

The owner should require monthly status reports that document work completed, the current schedule, and reason for any schedule changes, expenditures to date versus budget, and any project changes that were approved. Typical changes that occur on design–build projects include unforeseen site conditions, owner-specified value-added changes, and design errors and omissions that are not the responsibility of the owner.

Project change orders should be resolved and approved promptly by the owner and not be allowed to accumulate over the duration of the project. Procedures for addressing project scope changes should always be formalized as part of the contract negotiations with the design–build entity.

The owner should promptly provide for project inspection services as agreed to in the contract. The owner should schedule and attend regular progress meetings to enhance project coordination and information exchange.

TURNKEY CONSTRUCTION

Turnkey construction is typically used by private sector owners because only they are usually free to allow the large degree of control given the turnkey contractor. While turnkey provides the owner with a single point of responsibility as does the design–build concept, turnkey usually requires only that the owner define the concept of the project. The turnkey contractor specifies the requirements necessary to make the concept a reality.

Proper selection of an experienced turnkey contractor could greatly expedite project completion. Unlike the design–build concept, turnkey construction means the owner depends totally on the turnkey contractor from planning to design to construction to startup, and sometimes even operation.

Team Players and Responsibilities

The owner's responsibilities with a turnkey project are similar to those for a design–build project, except that the owner hires the contractor earlier in the project. The owner develops a conceptual plan for the project and the selection criteria for the contractor. The owner hires a contractor either through bidding or negotiating. Clear communication on the owner's part is crucial at the beginning so that the project gets started on the right path.

The owner designates a staff contact or hires a resident engineer or construction manager (CM). The CM serves as liaison between the owner and contractor and ensures project requirements are met. The CM reviews the design plan and specifications, performs on-site inspections, and approves equipment purchases. The CM must understand the owner's goals with the project and be able to facilitate construction, rather than trying to supervise the construction crews or otherwise unnecessarily obstruct the contractor.

The turnkey contractor must have the expertise to take the owner's conceptual plan and perform the planning (meeting platting, zoning, and legal building code requirements), develop site-grading plans, complete the preliminary and final design, prepare bid documents and hire subcontractors if necessary, and prepare equipment specifications. The contractor must be able to manage the construction and deliver a functional facility that meets the minimum requirements established by the preliminary plan and contract made with the owner. With some projects the contractor may train the owner's staff, or even in some cases operate the facility for a set period of time.

Contracting for Turnkey Services

Turnkey services can be contracted for in the same manner used for design–build. Because of the broad scope of a turnkey project, a turnkey project is often performed under a cost-plus-fee contract. Contractors for turnkey projects tend to be selected by negotiation rather than competitive bidding, particularly for private sector owners.

Effective Turnkey Procedures

Turnkey projects must be given considerable forethought and planning to be successful. The owner should take the following minimum steps before entering into a contract:

1. Prepare a preliminary (concept) plan for the project.
2. Prepare a project feasibility study and determine facility O&M cost.
3. If in-house project management expertise is not available, hire a technical advisor or construction manager. It is recommended that the CM be selected before selection of the contractor.
4. Select the turnkey contractor based on the contractors' proven experience in building similar facilities. Negotiate the fee based on the preliminary concept plan and the equipment requirements determined for the project. If services must be acquired through competitive bidding, use the concept plan to develop performance specifications and bid documents.
5. Require all designs and equipment submittals to be sealed by a registered professional and submitted to the owner for acceptance.
6. Include facility startup and training to be provided by the contractor as needed.

STANDARD DOCUMENTS

The Engineers Joint Contracts Document Committee (EJCDC) has developed standard construction documents for design–build projects, similar to the documents discussed in chapter 2. The Associated General Contractors of America (AGC) also has standard construction documents related to the design–build process. The following AGC documents are available as a reference for those who are considering design–build or turnkey. State or local contracting requirements should be reviewed before using these documents.

- Preliminary Design–Build Agreement (400). Includes a discussion of reimbursement of design expenses in the event the proposed project does not go forward.
- Design–Build Guidelines for Building Construction (405). Explains the advantages of design–build contracting and how to select a design–build contractor.
- Standard Form of Design–Build Agreement and General Conditions Between Owner and Contractor (Provides a Guaranteed Maximum Price) (410). Includes the terms of a formal agreement for design–build services for a guaranteed maximum price.
- Standard Form of Design–Build Agreement and General Conditions Between Owner and Contractor (Where the Basis of Compensation Is Lump Sum) (415). Similar to the AGC 410 but based on lump sum compensation.
- Standard Form of Agreement Between Contractor and Architect (420). For use by contractors engaged in the design–build method of contracting when contracting with an architect to provide design services.
- Conditions Between Contractor and Subcontractor for Design–Build (430). A companion to AGC 450 and AGC 450-1.
- Change Order/Contractor Fee Adjustment (440). For use by contractors in handling change orders on design–build projects or on any project in which a fee is involved.
- Standard Design–Build Subcontract Agreement With Subcontractor Not Providing Design (450). To be used with AGC 430 when the subcontractor does not perform design functions.
- Standard Design–Build Subcontract Agreement With Subcontractor Providing Design (450-1). To be used with AGC 430 when the subcontractor provides design services.
- Contractor's Reference Manual for Design–Build (490). This compilation of material includes contract documents relating to the design–build method of project delivery, as well as various guideline publications providing helpful information to owners and contractors. Included among the contract documents are AGC 400, 410, 415, 420, 430, 450, and 450-1. The guideline publications included are Design–Build Guidelines for Building Construction; Owner's Guide to Building Construction Contracting Methods; Guide to Owner's Responsibility for Construction Projects; Guidelines for the Contractor/Developer; and Guidelines for Obtaining Owner Financial Information.

REFERENCE

Siegel, D. 1995. Contract Documents May Ease Design–Build Process. *Engineering Times,* 17(5):11.

Chapter 10

Guaranteed Maximum Price Construction Management

With guaranteed maximum price construction management (GMPCM), the owner hires a construction management firm (CM) to perform contract administration. That firm guarantees a maximum price for the complete project. The owner and CM agree on the price before the construction phase of the project begins. Any cost overruns not due to unforeseen circumstances are borne by the construction management firm.

This chapter discusses the characteristics of a GMPCM project and the differences from other types of construction management, particularly CM-managed projects, as discussed in chapter 7. The responsibilities of the participants in a GMPCM project are also presented.

CHARACTERISTICS

Guaranteed maximum price construction management differs considerably from the more traditional CM role (discussed in chapter 7), where the construction manager is an agent of the owner and does not provide design or construction services. Instead, the owner retains the CM for its expertise in construction planning, coordination, contracting, and contract administration. With a GMPCM project, the CM assumes additional risk and usually has more direct involvement in the actual construction of the project. The CM must properly estimate the construction cost for the project based on the plans and specifications prepared by the design consultant. Although the owner of a GMPCM project still contracts directly with the contractor, the CM usually desires a voice in selecting the contractors. During the construction phase, the CM must schedule, coordinate, and control the construction process as much as in its power to keep costs within the guaranteed maximum price without compromising project quality.

The guaranteed maximum price does include contingencies or an amount for unforeseen circumstances. Changes in project size, materials, quality, site conditions, and schedule could result in renegotiating the guaranteed maximum price. Generally, the more specific the plans and specifications in terms of materials, quality, site conditions, and schedule, the fewer the changes needed. A CM arrangement can be changed to GMPCM when the project has been sufficiently defined so that a guaranteed maximum price can reasonably be provided. The CM must be willing and financially able to accept the additional risk.

Payment to the CM is usually either a lump sum or cost plus a fixed fee. The fee portion includes an amount for accepting the additional financial risk. Cost could include the construction manager's costs for labor, material, subcontracts, and incidentals, as allowed in the arrangement with the owner.

Variations

There are also some variations of a GMPCM project depending on what role the owner wants the construction manager to play (ASCE 1987). The construction manager can assume the role of "constructor" and construct a portion of the project with its own crews; the remaining work is constructed by other prime contractors that have a separate contract with the owner. Alternatively, the CM can perform some of the work and contract with other trades as a general contractor would. Or the CM does not perform the actual construction but acts as a general contractor that subcontracts directly with the prime trades. In all of these arrangements, the design consultant has a separate design services contract with the owner. An administrative or peer relationship exists between the design consultant and the construction manager.

Comparison to Design–Build

Despite a similarity to the GMPCM variations, the design–build approach is different from GMPCM. In the design–build arrangement, the design consultant is retained by or may be part of the design–build team's organization. The design–build team either designs and constructs the project entirely or directly contracts with the trades. Design–build is favored by owners that prefer a single point of responsibility for the project from design and construction to operation and performance.

Laws in some jurisdictions require that design–builders provide a warranty for the complete project. This requirement may increase the design consultant's liability above the traditional standard of care.

Comparison to Traditional Contracting

With traditional contracting, the owner contracts with a design consultant to prepare the construction contract documents. The project is bid, and the contract is executed with the lowest qualified bidder. The only changes to project cost occur when there are unforeseen site conditions; changes in scope, schedule, quality, or type of materials; or design errors or omissions. Under the traditional contracting arrangement, the final project cost is not known until construction is complete and all claims and change orders are settled.

The principal difference between traditional contracting and GMPCM projects is the presence of a CM that uses its expertise in construction planning, coordination, and contracting to define the construction budget and guarantee that the price to the owner will not be exceeded. The CM works with the owner and design consultant to arrive at the price, which differs from a traditional bidding phase. Based on the guaranteed price, the owner can determine whether the project is economically feasible.

OBJECTIVES OF GMPCM

The owner of any project has three objectives: conformance to expectations, economical cost, and on-time completion. Proper use of a CM in a GMPCM project can meet these objectives. The advantages and disadvantages of a GMPCM project can also be expressed in terms of these project objectives.

In general, projects that can benefit from the GMPCM approach are complex (Figures 10-1 and 10-2). The owner may not have the staff experience or expertise to complete the project. A qualified CM can provide that expertise and experience with a guaranteed maximum price.

Cost

Controlling the project cost is one of the principal benefits of a GMPCM project. In any project, opportunities for controlling cost diminish with time as the project proceeds from the planning stage through design and into construction. In GMPCM the CM is retained early in the project. The CM's expertise in construction, contracting, and estimating can influence decisions made by the design consultant to

Figure 10-1 Fort Myers, Fla., 12-mgd (0.5258-m^3/s) nanofiltration water treatment plant

Figure 10-2 Pump station (18.25-mgd [0.7996-m^3/s]) in suburban Washington, D.C.

keep project costs to a minimum in a manner consistent with the owner's objectives for quality and functionality.

These economic decisions generally fall in the area of material selection, contract packaging, staging, and sequencing. The owner must make sure that the CM and the design consultant communicate, generally through constructibility reviews. Cost management is achieved through multiple cycles of estimating, budgeting, and comparing throughout the design process. Of course, the CM must not usurp any of the design consultant's responsibility as a consultant in this entire process.

Once the project has been defined and developed, the CM guarantees the project cost at completion and manages the construction prime contractors to keep that cost under the guaranteed maximum. This ability to control and guarantee the costs up front before construction starts almost always results in lower project costs than are achievable under the traditional approach, even when the CM's additional fee for risk is included. The owner spreads some of the financial risk to the CM. However, the owner does rely on the ability of the CM to perform economically for the additional cost of the CM to be justified.

Quality

Quality of the project is expressed in terms of conforming to the owner's requirements as expressed in the plans, specifications, and other contract documents before establishing the guaranteed maximum price. Conformance to project requirements is monitored by the owner and the design consultant. Depending on the project organization and the owner's preference, full-time on-site representation (construction observation) can be provided by the design consultant or performed by a third party.

As with the project cost risk, the owner spreads more of the quality risk to the CM. In return, however, the owner does surrender some control of the project to the CM.

Figure 10-3 Foundation work completed (note silt fence in foreground)

Schedule

The greatest potential benefit for GMPCM is reduced project completion time through schedule control and management. Proper and realistic scheduling also ensures better cost control.

The overall project schedule can be further compressed using a fast track approach—common when the GMPCM arrangement is used. Under this approach, site work and foundation work (Figure 10-3) can start early and equipment items requiring long lead times can be purchased early. Design, procurement, and construction activities can overlap much more than in the traditional approach for which these activities tend to be sequential.

With GMPCM projects for which the CM performs some of the construction work or is responsible for contracting the work, construction time may be reduced somewhat, primarily as a result of reducing time required for bidding, bid evaluation, and awarding the contract.

PARTICIPANT RESPONSIBILITIES

Typical responsibilities of owner, design consultant, construction management firm, and contractor in each phase of a GMPCM project are discussed in the following paragraphs. Because of the wide range of GMPCM projects, responsibilities of the participants can also vary.

Planning

With a GMPCM project, the CM may be hired to help the owner plan the project, often before a design consultant is chosen. The following discussion assumes the owner hires a CM at the beginning of the project.

Owner. When the owner has identified the need for a project, it next decides how the project should be managed. As explained previously, many options are available. The owner must evaluate advantages and disadvantages of each option, as well as any relevant regulatory constraints.

If the GMPCM approach is chosen, the owner then usually selects the construction management firm. Selection is based on the firm's expertise and experience with similar projects. Because the project has yet to be designed, no CM could give a reasonable guaranteed price at this point. As detailed in previous chapters, selection criteria are similar for selecting a design consultant, a CM, and possibly a contractor (if the CM will not actually perform all of the work). Financial resources and backing of the CM in a GMPCM project are important factors in the selection process.

During the planning phase, the owner works with the CM to arrive at an optimal solution to meet the owner's needs. The owner's responsibilities include providing access to staff and facilities as needed. The owner must clearly communicate its needs and keep the project focused on those needs. The owner, who has hired the CM for its expertise and experience, should be receptive to the CM's solutions but must also make informed decisions.

The owner must clearly establish the responsibilities of the CM and design consultant. Compensation for the CM and consultant is determined by a fee structure as described in other chapters. The CM usually has a fee for the planning and design phase separate from the guaranteed maximum price for project construction.

Construction management firm. The CM must endeavor to fully understand the owner's needs, funding capabilities, and operations and maintenance (O&M) abilities once the project is completed. Only with this understanding can the CM help the owner and complete a successful project.

The general responsibilities of a CM in the planning phase can include

- clarifying owner needs
- identifying options to meet the needs
- estimating costs and funding possibilities
- helping the owner select the best option
- preparing reports

The CM may or may not advise the owner in selecting the design consultant. However, because the CM and the design consultant must work together as a team, the owner often involves the CM in the selection process. The CM should assist the owner in choosing a design consultant qualified and experienced both in the right type of project and the GMPCM approach.

Design consultant. The design consultant is most often hired during the planning phase to help the owner and CM develop alternatives for design and to select the best option. Bringing the design consultant in during this phase is usually advantageous; it reduces the time needed for final design.

The design consultant in a GMPCM project generally has more narrowly defined responsibilities than with other types of projects. For example, the CM usually estimates costs of design alternatives and has a greater voice in selecting the best option.

Design and Bidding

With a traditional project, the owner and design consultant are the two main participants in the design and bidding phases. With GMPCM, the construction management firm also participates and may actually manage procedures.

Owner. During the design phase of a GMPCM project, the owner's responsibilities are largely the same as with other types of projects. The main difference with a GMPCM project is that two entities other than the owner—usually the design consultant and the CM—are involved in the design. As with all phases of a project, the owner must clearly define the responsibilities of each participant before beginning work. The owner best assists the CM and consultant by being open, direct, and available for review, for approvals, and at other decision-making points in the process.

Construction management firm. The CM's main responsibility during design is to bring its construction management experience to the design process, so that the best possible design is reached consistent with the owner's budget. The owner may give the CM responsibilities during the design phase that the owner usually performs, such as managing the design consultant's work or conducting bidding.

The CM must clearly understand its authority and responsibilities for design and bidding. The CM must perform with the interests of the owner foremost in mind. There is a potential for disagreement between the CM and the design consultant because of different opinions on the design, and the CM is obliged to reduce that potential.

Whatever the specific GMPCM arrangement, the CM works closely with the consultant on design and must be available for review of the consultant's work on a timely basis. However, the CM should not manage the consultant's work and staff.

Often with GMPCM projects, as soon as the design is final, or even sooner if feasible, portions of the project such as site clearing or equipment ordering may begin. Of course, because the project is built at a guaranteed maximum price, that price will have to be agreed on before such work begins. Some risk exists in beginning the project before design is final, and the CM may build in additional contingencies in price to cover the unknown.

The CM and owner must also decide the extent of the CM's direct participation in construction work. This decision is usually based on the CM's expertise and experience. The CM usually handles the bidding phase of the project, using the plans, specifications, and contract conditions prepared by the design consultant.

Design consultant. Forms of the contract documents (drawings, specifications, contract conditions) prepared by the design consultant under the GMPCM approach are not much different from those prepared under the traditional approach. It is important to indicate that the GMPCM approach will be used, so that the general contractor or prime contractors understand and recognize the role and responsibility of the CM.

If the CM will construct all or part of the work by retaining prime contractors directly, the contract documents are "packaged" differently—generally by means of trade subcontracts. The design consultant must work closely with the CM to identify individual contracts to ensure there are no gaps in work. Any such gaps might slow work later in the project. Under the GMPCM arrangement, the CM would be liable financially in such a case.

During the design phase, after the guaranteed maximum price has been established, consultants sometimes try to perfect or better the design, for example, by specifying slightly better-quality material or perhaps specifying that a concrete slab be another inch thick to be conservative. Unless the initial decision is incorrect, the CM usually resists these types of changes; obviously the guaranteed price must be based on a set final design. Project expenses can become higher than expected as a result of such changes in specifications, which makes it difficult for the CM to control the budget.

However, after the guaranteed maximum price has been established, if better materials or methods are discovered that will increase the project "value" or reduce long-term maintenance or operating costs, it is the duty of the design consultant to bring it to the CM's attention. The CM should then secure the owner's approval with an appropriate increase in the guaranteed maximum price for the project.

If the project is on a fast track, the design consultant might have to work with decisions made early on the project, perhaps even before the design consultant was hired. This situation often creates difficulties because the designer may have to work with less-than-perfect decisions or have the owner and CM agree to changes in these decisions. The owner and CM must weigh options, which usually come down to a choice between spending more money or more time.

Construction

As with other phases, the construction phase responsibilities in GMPCM projects vary according to contractual arrangements. In particular, the CM's duties vary from project to project. The following discussion presents typical responsibilities for each of the main participants. Major differences from traditional and other nontraditional projects are also presented.

Owner. The owner of a GMPCM project relies heavily on the CM to represent its interests during the construction phase. However, the owner is ultimately responsible for both quality and cost of the project and performs several duties, such as

- resolving field problems
- performing inspections as required
- reviewing and approving changes
- processing pay requests
- attending construction progress meetings

Construction management firm. The construction management firm is broadly responsible for the quality and cost of the project. The CM may be more involved in actual construction of the project, depending on the role the owner wants the CM to play. Specific responsibilities vary considerably from project to project. As always, these responsibilities must be clearly established and agreed to by all participants.

Important responsibilities that must be established include

- inspections and materials testing coordination
- construction monitoring and management
- shop-drawing review
- field authority
- design-change review
- progress-payment approvals
- permit and environmental compliance monitoring
- community relations

The CM has a site representative (resident manager or project manager) who is always at the project and who is responsible for coordination, communication, and scheduling.

Design consultant. The design consultant in a GMPCM project usually does not perform construction management, but it does perform other typical tasks required of the designer including

- shop-drawing review
- design change
- field problem resolution

The design consultant rarely has a site representative on a GMPCM project; however, the designer should have a staff contact available.

Contractor. The general contractor of a GMPCM project may or may not be the construction management firm. If not, the contractor's responsibilities depend on which contracting approach is selected, as discussed in earlier chapters.

Figure 10-4 Bank of feed pumps

Postconstruction and Operations and Maintenance

The contract between owner and CM in a GMPCM project should clearly state the responsibilities of all participants in the postconstruction phase. A project is rarely considered complete when installation is finished. Startup, testing, training, and other tasks are very important to the success of the project. The CM must include in the guaranteed maximum price the costs of these tasks.

Owner. The owner typically makes its staff available to participate in the testing and startup, prepares a punch list of unfinished tasks, and performs final inspections. The owner's staff is also trained in O&M procedures by the CM, the contractor, the equipment suppliers, or the design consultant.

Construction management firm. The CM usually coordinates startup and testing. For a typical treatment facility this can include

- constructing flow bypass systems and recirculation systems for equipment testing
- obtaining chemicals for testing chemical feed systems (Figure 10-4)
- arranging for disposal of test fluids
- testing alarms, variable-speed equipment, and mechanical, electrical, and instrumentation equipment

When equipment is thus checked, it must be started up and operated for a specified period of time before it can be considered acceptable. The owner and design consultant may periodically observe performance during this testing and validation period. Some design consultants specify this validation period should be as long as 30 full days.

The owner has final approval before a facility is put into actual operation. The CM is usually responsible for assembling facility O&M manuals and equipment user manuals from suppliers.

Design consultant. The design consultant typically has fewer responsibilities during the postconstruction phase of a GMPCM project. The CM, as already discussed, handles coordination and scheduling. However, the designer typically observes testing and startup, may provide specialized training, and is called on when a question or problem arises. The designer may certify to the owner that the facility is performing to design specifications.

Contractors. Contractors, especially equipment suppliers, usually perform testing and startup procedures. They also perform adjustments as needed to meet performance requirements.

REFERENCE

American Society of Civil Engineers (ASCE), Committee on Construction Management. 1987. Qualification and Selection of Construction Managers with Suggested Guidelines for Selection Process. *Civil Engineering,* 113(1).

Section 3

General Issues

Partnering

Communication

Disputes

Payment Procedures

Private-Sector Financing

Chapter 11

Partnering

Partnering is a management method in which all of the participants make a special effort to cooperate in working toward completing the project successfully. Partnering can lead to many benefits, including: improved morale, reduced costs, development of innovative solutions, and effective problem solving or troubleshooting.

As defined by the US Army Corps of Engineers (Pamphlet 4, *Partnering,* IWR Pamphlet-91-ADR-P-4), partnering is an agreement in principle by the participants to share the risks involved in completing the project and to establish and promote a mutually supportive environment. Partnering is not a contractual agreement, nor does it create any legally enforceable rights or duties. Rather, partnering seeks to create a new cooperative attitude in completing contracts. To join in this partnership successfully, each member must seek to understand the goals, objectives, and needs of the other parties and to mutually seek a working relationship with other members.

Critical to successful partnering is the creation of an owner–contractor relationship that promotes the achievement of mutually beneficial goals. This approach is useful with a single prime contractor, but it works particularly well with a multiple prime contracts project (chapter 8). Close association among the owner, design consultant, and construction manager (CM) through the entire construction process is of great benefit to the owner and to team members. Good specialty contractors work effectively in the team approach under owner–designer–CM direction.

Figure 11-1 shows a signed partnering agreement for a plant expansion. The goals and objectives agreed to by the project participants are listed.

The key players in a typical partnering scenario include

- owner
- engineer
- general contractor
- electrical contractor
- heating, ventilation, air conditioning (HVAC) contractor
- plumbing contractor
- other prime contractors

Nansemond Plant Expansion

Phase 1

We are committed to the successful completion of the Nansemond Project meeting the goals and objectives of all parties through effective communication and mutual cooperation.

Our goals and objectives are as follows:

- Utilize and maintain a reliable schedule.
- Provide a quality facility.
- Achieve effective and timely resolution of job issues at the lowest possible level.
- Provide a safe work environment.
- Meet all environmental requirements.
- Respect financial objectives of all parties.
- Accomplish smooth transition, including startup and successful project closeout.

In the spirit of cooperation, teamwork and understanding, we commit to earn mutual respect and trust, which will result in a sense of pride and enjoyment of all participants.

Figure 11-1 Partnering agreement

- equipment suppliers
- material suppliers

A key to making partnering successful is effective communication of needed information, particularly that from the owner. For example, the owner knows treatment plant characteristics, is aware of different ways to handle existing flows, and is familiar with the underground distribution system. Project contractors need this information to perform their work, and the owner should willingly provide it.

The presence of an experienced resident inspector, either from the owner's or construction manager's staff, is also important to successful partnering. When a resident inspector is not on the project, problems are not solved as quickly. The project engineer's responsibilities in partnering include reviewing shop drawings, answering technical questions, and sharing expertise with the particular type of project. The general or trade contractor in a partnering scenario should work closely and openly with the owner and engineer, and vice versa. In traditional construction projects, the contractor can feel a sense of "us against them," since its work is constantly being scrutinized and it is pressured to complete the work on time.

Partnering applies to all phases of the construction project and should not be used only when convenient. For example, toward the end of a project, it is easy for the participants to go their separate ways. However, if the owner, the engineer, equipment suppliers, and contractor continue to work together at the time of equipment startup, the process will go more smoothly, and the project can be brought to a more satisfying conclusion.

CASE STUDIES

The following case studies illustrate problems encountered when team members do not work together, as well as benefits gained when they do.

Partnering Case Study 1

The first case study involves two, 2-m belt filter presses. In the proposal, the engineer left the polymer specification open, leaving the choice of polymer feed equipment up to the contractor. The filter press equipment supplier ran a pilot test at the plant site to predict how the equipment would perform under full-scale conditions.

At project completion, the belt presses did not perform as specified—the 50 percent sludge mixture specified in the contract was not achieved. The contractor and equipment supplier were facing a large penalty for additional costs to the project.

Two main factors caused the belt presses not to perform as specified:

1. The polymer equipment furnished by the contractor was not compatible with the belt press equipment requirements but was approved by the engineer.
2. The belt press equipment had mechanical problems.

Under such circumstances, the project can go in one of two directions:

1. No one agrees to a solution and large penalties are assessed, often resulting in litigation to determine blame.
2. The parties put aside their differences to create an appropriate solution.

In this particular situation, all parties acknowledged that there was fault on everyone's part, and all parties worked together to solve the problem. Soon, the belt presses were operating satisfactorily, in fact, better than specified. Liquidation damages were not charged because there was full cooperation from all parties. The result was an excellent installation, with all parties satisfied they had cooperated fully toward fulfillment of their common goal.

Partnering Case Study 2

The second case study involves a project in which a brand and model of equipment was specified as the only option. The contractor questioned the engineer and owner before the bid to confirm that this was a sole source specification. After bids were received, the low-bidding contractor approached the owner with an offer to reduce the contract price considerably if allowed to use alternative equipment. The proposed equipment, in addition to being less expensive, had a proven track record.

The owner disagreed and instead wanted to use the equipment specified in the contract, despite the fact that there had never been an acceptable installation of this equipment. In fact, this project would be only the second in production for this equipment. Early in the construction phase it became obvious there would be many problems with the equipment. The highly sophisticated equipment, with many electrical components and computerized operating parts, was being installed in the open without cover. Heat, cold, rain, dirt, and other contaminants damaged the equipment. A cover had originally been included in project specifications, but the first round of bids came in too high, and the specifications were revised. The protective cover would have provided protection for the computer, air compressor and components, air valves, and electrical controls. The cover was installed later under another contract.

Further complicating the situation, the owner would not allow its personnel to work with the equipment until the project was complete. Earlier testing of the equipment would have helped to identify some major problems, such as a problem with the incoming electrical power. The supplier eventually resolved the power problems, though they were not the supplier's fault. Upon completion of the project, the owner withheld a large amount of payment due the contractor because of the equipment problems and the increased costs to solve them.

A partnering approach could have led to lower costs and a much more satisfactory conclusion to this project:

- The owner could have given more consideration to the contractor's recommendation to use proven equipment.
- The owner could have encouraged staff to work closely with the equipment supplier during installation. The power supply problem would have been discovered earlier, and the staff could have made additional provisions for protecting the sophisticated equipment.
- The supplier would not have been penalized for problems it did not create.

Partnering Case Study 3

Another example is a project in which a deteriorating timber pile foundation threatened to shut down a major West Coast grain export facility. In this project, the goal was to introduce high-capacity pin piles to support collapsing grain silos. Paul B. Groneck (1993) writes that several factors complicated the project:

- Operations of the silos could not be interrupted.
- Access was extremely limited for work in the basements of the silos.
- The explosive nature of the grain dust prohibited use of combustion engines and any metal welding, cutting, or grinding equipment in the basements.
- Extending the underpinning elements among the existing timber piles—which were very competent below the top of the water table—down to the competent bearing level of the piles would be difficult.
- The concentration of the foundation support around the columns and walls could have overstressed the minimally reinforced concrete walls and slabs.

This project required innovation and careful planning. Special drilling techniques were employed that limited the problems caused by the high density of the lower gravel and cobble soils and the maze of timber pilings.

Construction began in February 1992 and was completed in January 1993. On March 25 of the same year, an earthquake measuring 5.6 on the Richter scale shook the site. With the epicenter just 44 miles away from the construction site, extensive damage could have been expected. However, none occurred, demonstrating the quality of the construction.

Groneck attributes the success of the project in a large part to effective partnering efforts. At the start of the project, all parties signed a one-page agreement stating common goals of safety, successful and timely completion, uninterrupted operation of the facility, and maximization of the contractor's profit. This agreement helped establish positive attitudes and an open level of communication.

Partnering Workshops

Principles of partnering are simple enough to understand. However, implementing them may require changing attitudes and behaviors—which is not so easy to do. A partnering workshop might be helpful for the participants who are not familiar with the concept's principles. The workshop should be facilitated by someone who has extensive experience in partnering. Figure 11-2 shows a typical agenda for a two-day partnering workshop.

Day One

7:45 a.m.	■ Registration and continental breakfast	
8:00 a.m.	■ Welcome and administrative announcements	— Facilitator
	■ Introductions, organization overview, expectations for the workshop	— Participants
	■ Baseline evaluation of working relationships	— Facilitator
	■ Video ("The Business of Paradigms") and discussion (the acceptance and possible benefits of change)	
10:00 a.m.	■ Break	
10:15 a.m.	■ The partnering concept	— Facilitator
	■ Small group team-building exercise (teamwork, trust, and communication)	— Facilitator
12:00 p.m.	■ Lunch	
1:00 p.m.	■ Develop project team's partnering pledge (large and small groups)	— Facilitator
2:30 p.m.	■ The partnering process, ground rules, and the problem-solving process	— Facilitator
3:30 p.m.	■ Break	
3:45 p.m.	■ Large-group brainstorming; identifying, combining, and prioritizing issues and concerns	
5:30 p.m.	■ Social hour	
6:30 p.m.	■ Dinner	
7:30 p.m.	■ Evening program	

Day Two

7:45 a.m.	■ Continental breakfast	
8:00 a.m.	■ Review issues and concerns previously defined	— Facilitator
	■ The definition and action plan process	— Facilitator
	■ Small groups define assigned issues and concerns and develop action plans	
	■ Report to large group, develop consensus	
10:30 a.m.	■ Break	
10:45 a.m.	■ Small groups define assigned issues and concerns and develop action plans	
	■ Report to large group, develop consensus	
12:30 p.m.	■ Lunch	
1:30 p.m.	■ Small groups define assigned issues and concerns and develop action plans	
	■ Report to large group, develop consensus	
3:30 p.m	■ Break	
3:45 p.m.	■ Select and commission focus group	— Facilitator
	■ Review workshop achievements and commitments	
	■ Wrap up	
5:00 p.m.	■ Adjourn	

Figure 11-2 Typical agenda for a partnering workshop

REFERENCE

Groneck, P.B., D.A. Bruce, J.H. Greenman, and G. Bingham. 1993. Pin Piles Save Silos. *Civil Engineering*, (9):66-8.

Chapter 12

Communication

Throughout this manual, direct and open communication has been mentioned as being critical to successful project management. Good communication increases the chances of project success, but of course it does not happen accidentally. When project participants understand the fundamentals of the communication process and conscientiously apply them, there is a good chance communication will be effective.

This chapter presents the basics of good communication, describes the types of communication, and gives guidelines for effective project communication.

BASICS OF GOOD COMMUNICATION

The process of communicating can be presented in a simple model (Figure 12-1). In this model, two participants are a "sender" and a "receiver." The sender sends a message, which the receiver tries to understand. Interfering with understanding is "noise." The most important part of the model is the receiver. A sender needs to understand the receiver—his or her technical knowledge and needs—before communicating. In other words, a sender should be sensitive to the receiver.

Usually, noise is thought of as interfering sounds, such as the noise of construction machinery. That noise would interfere with two people trying to talk to each other. Noise can come from a variety of other sources, such as unclear writing, anger, poor photocopies, lack of attention on the receiver's part, or misspelled words.

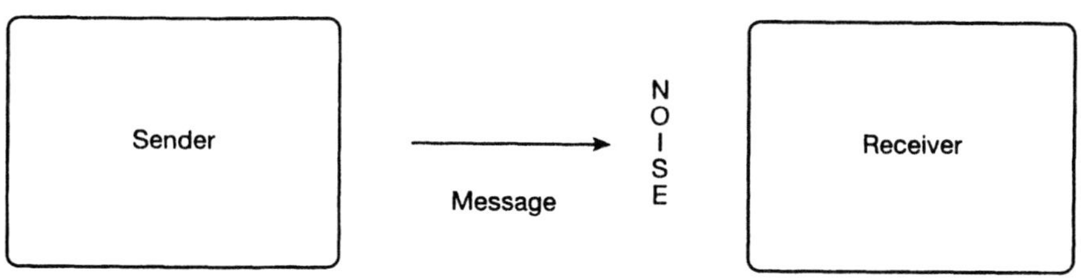

Figure 12-1 Communication model

Figure 12-2 Construction signage displaying a concise message

Eliminating noise

Eliminating noise is a key to effective communication. A message that has eliminated noise is

- clear
- concise
- specific

In a clear message the sender has made the message as understandable as possible. The sender uses words that the receiver understands, not words understood only in the sender's profession. For example, an engineer talking to homeowners about a treatment plant expansion should clarify terms such as *flocculation*. In addition, the sender's message should be unobstructed by physical noise or other "noise" such as anger or mistrust.

A concise message starts with the sender determining exactly what information—the topic—should be communicated. The message should then be restricted to the topic and should not obscure the important information. Nor should important information be left out. For example, does the homeowner's group need to hear a technical lecture on flocculation? The group may be interested only in the expansion's effect on rates. Figure 12-2 shows an example of a concise message.

Both a sender's writing and speaking style should be concise. For example, look at the two sentences in Figure 12-3 and decide which is "noisy" and which is concise.

A specific message uses concrete rather than abstract words. Abstract words are open to interpretation, whereas concrete words are less so. In the example in Figure 12-3 the contractor must give at least 48 hours notice, not "due" notice or "adequate" notice.

Anger

Trying to communicate in anger is one of the most disruptive sources of noise. Unfortunately, construction projects can be highly stressful to the participants,

It is vitally important that the owner be notified by the contractor with due notice before the procedures of the startup process are initiated.	The contractor must notify the owner at least 48 hours before startup.

Figure 12-3 Contrast in communication

and that stress often overflows into anger. The source of the anger is often frustration based on participants' lack of understanding of each other's situations. For example, suppose the contractor has bid the project as tight as possible and is worried about losing money on the job. The owner has to answer to the utility's board members, who question every penny being spent. The design engineer worries about the chance that the calculations were based on inaccurate field data. These different concerns cause each person to approach a problem differently.

The key to reducing anger, besides remaining calm, is first to seek to understand the other's position. Often the disagreement is not one of facts but interpretation. A contentious issue can also be divided into subissues, some of which can be agreed to by all. Finding out what the participants can agree on will often diffuse anger.

If the situation has reached the boiling point, it helps to back off until anger has cooled. It is important not to say things that will require an apology later and not to bring up past disagreements or issues that have nothing to do with the current problem. In dealing with anger, it also helps to "detach" oneself from the situation; in other words, it helps to reduce one's personal involvement by acting as a mediator in one's own disagreement. One must remain objective and not become defensive.

Purpose of the Message

An effective message has a purpose. Knowing that purpose allows a sender to focus on the message. The main purposes of communication are

- to inform or clarify
- to instruct
- to document
- to persuade

To inform or clarify means that the sender is imparting knowledge that the receiver desires. Before sending the message, the sender should determine what information the receiver needs and how best to communicate that information without noise. For example, most memos or phone calls are messages that inform or clarify. The sender has information or the receiver requests information, such as test results or an interpretation of a specification. The sender informs the receiver of the test results or gives an interpretation. Messages that inform or clarify should be concise.

A message that instructs is intended to teach the receiver a concept or procedure. Manufacturer's equipment manuals or a training session are examples. Instructional messages, particularly those for procedures, are best presented as concise procedures listed step by step in the order they need to be completed.

Document messages provide a permanent record of decisions, events, or results. Plans and specifications document the results of the design process. Meeting minutes document the discussion and decisions made at a meeting. Contracts document an agreement between two parties. Construction diaries document the daily events on the project. Document messages should be clear and unambiguous.

Messages that persuade try to convince the receiver to accept the sender's position on an issue. For example, by making business proposals, a firm is trying to be selected to perform consulting services. Persuasion is also used, for example, to convince others that a particular treatment option or piece of equipment is better. Clear, concise facts, as opposed to unsubstantiated opinions, are the most persuasive.

Managing Meetings

Meetings can be an effective way of communicating and reaching understanding, but they can also be ineffective. The difference is that effective meetings are managed. Just as effective communication is no accident, good meetings need good techniques. Most effective meeting management styles take the following issues into account:

- *Purpose.* Have a clear purpose for the meeting. Whether for a regularly scheduled progress meeting or a special meeting called to address specific issues, make sure the participants know the purpose so they can prepare.
- *Start and end times.* The most important meeting management step is to set start and end times and stick with them.
- *Agenda.* Send each participant an agenda before the meeting that lists the purpose of the meeting, the start and end time, the place of the meeting, the participants, the items to be discussed in the meeting, and the proposed time limits for each.
- *Focus.* Keep the meeting focused on the agenda items and on reaching agreement or arriving at required actions. Do not let the discussion get off course or be dominated by one participant. Elicit the input of all members. Use communication basics to keep the discussion focused.
- *Summary and minutes.* When an item has been completed, give a summary of the agreement or the required action and deadline to make sure everyone understands. Record the summary as part of the minutes.
- *Next meeting.* If needed, set the time and place for the next meeting. Send a copy of the minutes to each participant before the next meeting.

PROJECT COMMUNICATION

During a construction project, many communication events will occur. The typical communication events and messages in each phase of a project are discussed next.

Planning

During the planning phase, several meetings usually take place to define the scope of the project, establish priorities and budgets, select project team members, and develop a plan of action. These meetings usually involve only the owner's staff. It is important to keep good summary minutes so that nothing has to be discussed again. It is rarely necessary to keep detailed records of the discussion leading up to agreements.

If a consultant is hired to help with planning, the owner should make clear what types of communication the owner needs and at what points in the process. Usually there is an initial meeting at which the consultant meets the owner's project staff and clarifies the project scope. The owner and consultant should decide who will be the meeting manager and take minutes.

The consultant then usually works with individual staff members to develop alternative design or other project criteria. These communications are done in one-on-one meetings or telephone calls as necessary. There is usually a meeting to present alternatives and other items in a preliminary report to the owner's final decision maker. A final report and presentation are then prepared based on feedback from authorized reviewers to the preliminary report.

Design and Bidding

The design process itself is similar to the planning phase, with several meetings held to solve problems or present options. Meetings serve to refine the project scope and to define the quality criteria the owner desires (and can afford). The key to these meetings is to make sure all those involved understand the agreements reached on design issues. This understanding will reduce many problems later.

Results of the design process are documented in plans, specifications, bidding, and contract documents. Above all, these documents must be clear and specific. Previous chapters have discussed the specifics of the bidding process, including having a prebid meeting to clarify the plans and specifications and contract documents, as well as sending addenda to the bidders as needed. Contract documents were discussed in chapter 2.

Construction

The construction phase is filled with meetings and documentation. The usual meetings include

- a preconstruction meeting, which includes all participants as well as any other utilities or agencies affected by the project
- regular progress meetings, which usually include the contractor, construction manager, owner, and other parties immediately affected by the current work
- problem-solving meetings, which include the affected parties

The basics of good meetings also apply to construction meetings, in particular keeping them focused and starting and ending them on time. Otherwise, the meetings will be perceived as a nuisance rather than an effective method of keeping the project on course.

Construction documentation has also been discussed in previous chapters. Examples of such documentation include pay requests (also discussed in chapter 14), change orders, construction manager's diaries, and test results. Again, the key to good documentation is making it clear and specific.

Operations and Maintenance

After construction is complete, the main activities are related to operations and maintenance, in particular manuals and training. These activities are instructional communication. The key to good instruction is presenting the information in clear, chronological steps. Keeping the terminology understandable is also important.

Chapter 13

Disputes

The complexities of the construction process are fertile ground for disagreements and misunderstanding which, if not promptly resolved, can grow to the level of a dispute. In the past 20 years the number of disputes among various parties of the construction contract have risen dramatically. The number of attorneys and legal firms with large divisions that specialize in construction claims and disputes has also risen. Disputes can lead to time-consuming, costly resolution, often through litigation. The cost impact is heavy on all parties, even the prevailing party.

Because of the high cost of dispute resolution, more emphasis is now being placed on claims avoidance and management. Earlier chapters have emphasized the importance of proper project planning, design, contracts, and administration in minimizing construction claims and disputes. Nevertheless, disputes can occur even with the best efforts.

This chapter defines terms and common causes of disputes. It also presents methods of resolving disputes, including negotiation, arbitration, litigation, and alternative dispute resolution such as the use of mediation, a summary jury trial, or a dispute review board. The focus is on providing the contracting parties insight into the various dispute resolution methods and describing pitfalls that lead to problems, claims, and disputes.

BACKGROUND

The following discussion defines terms used in dispute resolution and identifies common causes of disputes.

Definitions

Problems during construction are not unusual; in fact, the entire design and construction process consists of problem resolution. All the parties—owner, engineer, contractor, subcontractors, suppliers, and others—involved in the project undertake a certain amount of risk. When a problem goes beyond the risks initially accepted by the parties, a dispute often arises. A construction problem becomes a claim when a party asks for more money or time or both that result in no tangible benefit to the owner.

For example, a contractor may view site conditions that differ from those originally anticipated as a basis for making a claim, but the owner might not consider the claim justified. The contractor can then make a formal claim to offer sufficient justification for all or a portion of the claim, and a settlement can be negotiated on site. A change order then formalizes a contract change, and the issue is resolved. A change order is a response to a problem that is agreed to by all parties. It is an agreed change to the contract and can call for an increase or decrease in both price and time.

A claim escalates to a dispute when the other parties cannot agree on the justification for the claim. Proper claim procedures involve the presentation of a well-documented claim that advances the merits of the claim, together with documentation on the associated cost, schedule, and other particulars. The party to whom the claim is presented must make a thorough and fair evaluation. Unfortunately, this is rarely the case. Typically each of the parties advances to an uncompromising position without really knowing the value of their position.

Causes of Construction Disputes

The causes that can lead to a construction dispute are numerous and often complex. Disputes are often the result of an accumulation of minor confrontations and disagreements. The increasingly litigious nature of the industry causes dispute participants to direct much energy toward "building a case" and taking a position from which they will not move for fear of compromising their case. This makes it important for the various parties to understand what can be involved in resolving a claim once it has reached the dispute stage.

Two factors equally important in understanding why disputes arise are the origin of the underlying problem and the way parties handle the problems. Insight into some of the major causes of disputes can be helpful in avoiding and resolving disagreements before they become full-blown disputes.

Thirty or 40 years ago the perception of the construction contracting process was entirely different from today. Relatively few entities were involved other than the owner and a contractor. Today, many more entities and experts are directly involved in the project, such as consultants and subcontractors. In addition, other entities outside the contract can exert considerable influence on the project, such as regulatory agencies, local government units, state highway departments, the Army Corps of Engineers, environmental agencies and organizations, and neighborhood and citizens groups.

Figures 13-1 through 13-4 illustrate some of the typical contractual arrangements that are encountered in water utility projects. Figure 13-1 depicts the originally conceived form of contract; the owner contracted with an engineer for design and made a separate contract with the general contractor to build the facility.

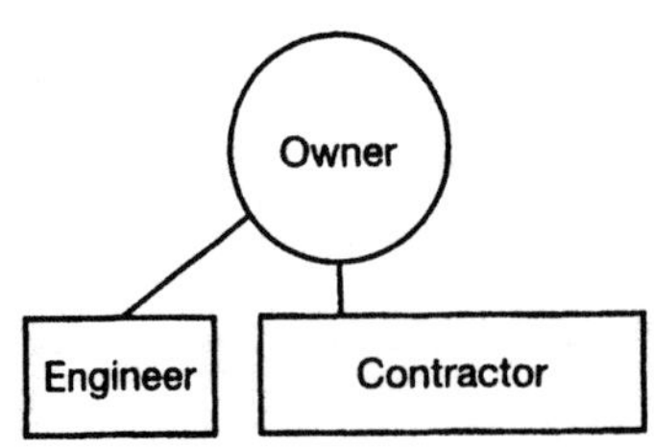

Figure 13-1 Typical contractual arrangement 30 years ago

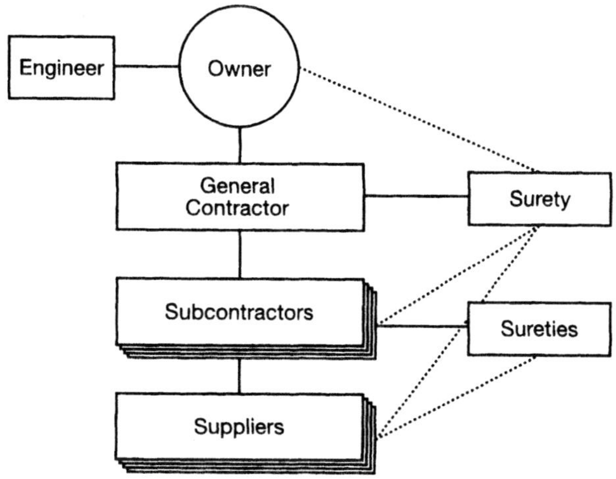

Figure 13-2 Typical general contractor format today

Figures 13-2 to 13-4 illustrate the much more complex contractual process represented today. The term *surety* is the guarantee that the various parties will perform, and may take the form of performance bonds or payment bonds. All of the entities identified in these contractual relationships are potential dispute adversaries. It should be well recognized that the many participants have different roles, policies, and operating procedures, as well as different personalities and behavior. The variety brought to the process is essential in this age of rapidly changing technology and specialization. However, the very complexity of the player matrix creates an environment that is more conducive to misunderstandings that lead to disputes. All of this definitely suggests that the contract process that ties various entities to the project must be carefully crafted.

The following paragraphs discuss some of the specific causes of disputes.

Contracts. Because of the complexities noted previously, combined with the complexities of the particular project, the construction contract itself can be the main source of problems contributing to disputes. Some of the problem areas with contracts include ambiguous contract language, outright omissions, conflicting requirements, and unrealistic time or performance requirements.

Changes. When the contractor is required to undertake any type of work that deviates from the original bargain of the contract or from the scheduled work plan or time of performance, a change in the project occurs. A dispute typically arises when the contractor seeks adjustments that the owner feels are excessive.

General classes of changes include formal changes, constructive changes, and cardinal changes. Formal changes are the traditional change orders that direct specific changes, acknowledge the change is being made, and involve a formal contract change-order provision. Constructive changes result from actions of the owner or its agents that have the effect of a change directive but often without any formal recognition of a change. Cardinal change occurs when the change is sufficient to change the entire character of the work.

Regardless of the reason or justification, changes often have a ripple effect that impacts other areas of work and escalates problems that were difficult to foresee at the time the change was started.

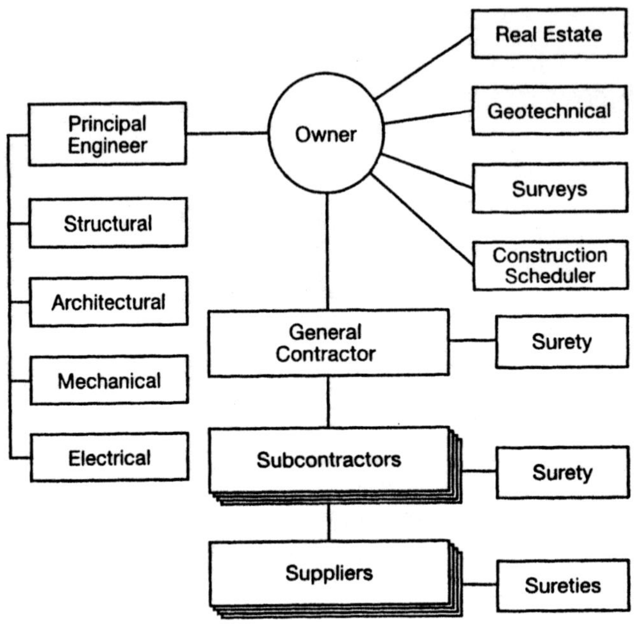

Figure 13-3 Typical format for modern large projects

Site conditions. Different site conditions from those predicted are perhaps the most frequent reason for a claim that leads to a dispute. In many cases, the owner has attempted to shift the risk of the unknown onto the contractor in an attempt to reduce up-front design or investigative costs. However, the contractor that must win by making the lowest competitive bid cannot afford to include any substantial amounts of contingency costs. Disputes arise over differing opinions as to what should "reasonably" have been expected.

Delays. A delay more often than not impacts costs for all parties involved in the construction process. A delay can be caused by any one or combination of the parties. The owner's failure to obtain timely financing and rights-of-way, give approvals, or render decisions can lead to delays that result in claims for extra compensation. Likewise, a supplier's untimely delivery of materials can delay other subcontractors and ultimately delay the entire project, adversely affecting the owner and the engineer financially. Frequently the engineer handles the construction-phase services on a limited budget; delays on the project lead to claims for extra compensation. The owner, more often than not, is reluctant to recognize the full cost impact of delays on the other parties.

Quality or performance. Poor installation, defective materials or equipment, and even construction failures are typical problem areas for disputes. Many of these problems may not become apparent until well into the construction process or even after the project is completed and put into operation. Such problems discovered at this stage are usually very difficult to remedy to the satisfaction of all parties.

Outside forces. Forces outside the reasonable control of any of the contracting parties may dramatically impact the progress and cost of a construction project and result in a claim. Such naturally occurring forces include floods, earthquakes, wind, and unusually severe weather. Other outside forces of a less dramatic nature are more common concerns. Intervention by a governmental agency, a public body, or

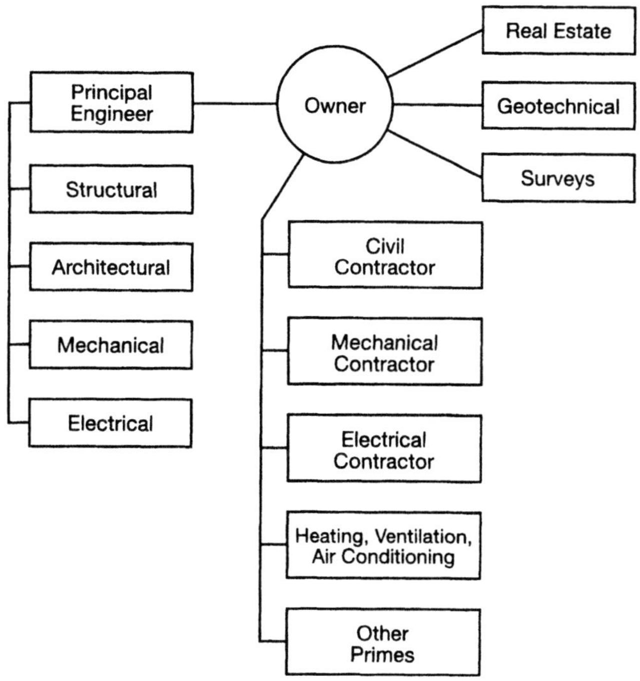

Figure 13-4 Typical modern format with multiple prime contractors

citizens group can impede, change, or delay a project. Adequate planning and anticipation can reduce the likelihood that these forces will cause problems.

DISPUTE RESOLUTION

The three traditional methods of dispute resolution are

- negotiation
- arbitration
- litigation

More recently a fourth method has evolved—alternative dispute resolution. Regardless of which method is chosen, several steps should be followed for a successful resolution, as discussed next.

Preparation

Once it becomes apparent that a dispute must be resolved in a formal manner, experts in the area of construction claims and litigation should be consulted. Over the past 25 years, the area of construction claims litigation has become a relatively complex and specialized field. Even those owners and their engineers who have a capable staff may need to be supplemented by experts who can render at least a knowledgeable second opinion on the basis of the claims. Detailing all of the preparation needed to resolve a dispute is impossible, but the following general preparatory steps are most important:

- Identify and isolate the problems responsible for the dispute, including personality conflicts and emotions.

- Organize all available facts, documents, notes, and circumstances that bear on the problem.
- Give proper and timely notices to those entitled to receive them.
- Determine what and to what extent additional assistance from consultants, or others may be required. For example, a scheduling consultant may be required for analysis of delay claims.
- Determine options, probable costs, whether a settlement is possible, whether there are grounds to sue, and whether contract documents provide for specific means of dispute resolution.

Negotiation

A negotiated settlement should almost always be considered as the method to resolve a dispute. The art of negotiation is always subject to the particular style and experience of the participants. However, some of the following elements are common to successful negotiations.

Be prepared. It is important to know the details and anticipate the opposing party's arguments. The relevant facts should be presented and documented with letters, memos, logs, and contract provisions. Charts, graphs, drawings, and photography can be used effectively to illustrate points. The cost components or specific disagreements should be identified with respect to the claimed costs. A comparison of "as planned" versus "actual" scheduled tasks in a chart form is always helpful.

Pick the negotiating team. Different circumstances call for different types of negotiators, but two types of individuals are usually essential for successful negotiation: the party's representative most closely associated with the project, and the party's decision maker. The party's attorneys may also be present.

Know negotiation techniques. Several books and manuals on negotiation are available. An understanding and appreciation of the opponent's particular problems that contributed to the dispute will be helpful in forming a negotiating strategy.

Prepare a memorandum of agreement. The results of the agreement reached through the negotiation process should be promptly summarized in a memorandum of agreement. A complete record should be made that identifies the important documents and points presented, the presenting party, and the times presented.

Arbitration

If attempts at negotiation prove fruitless, the claimant may have two alternatives: arbitration or litigation. The availability of arbitration as an alternative depends on whether or not the contract contains an enforceable arbitration clause covering the dispute.

Today, contract clauses providing for arbitration of disputes are commonplace. The Engineers Joint Contract Documents Committee (EJCDC) Standard Forms of Agreement, typically used in water utility projects, provide for arbitration in accordance with the Construction Industry Rules of the American Arbitration Association (AAA). Arbitration is provided for in the EJCDC standard contract between owner and engineer and between the owner and contractor.

An important provision of the arbitration clause prohibits the inclusion of another entity that is not a party of the specific contract. This means that arbitration between the owner and contractor cannot include the engineer or other parties in the proceedings.

Following the AAA rules, the complaining party files a Demand for Arbitration with the AAA regional office. The Demand for Arbitration does the following:

- names the parties
- describes the nature of the dispute
- requests it be resolved by arbitration
- recites the agreement to arbitrate
- designates the location for requested arbitration proceedings

The demand need not be formal nor detailed, but it must be sufficient to notify the AAA and the adversary of the foregoing elements.

The AAA administers the arbitration, assigning it an identifying number and notifying the adversary party of the demand. The adversary then normally has one week in which to file its answer to the Demand for Arbitration, failing which the allegations contained in the demand are automatically denied. (There is no award by default in arbitration proceedings.)

If the adversary files an answer, the AAA then furnishes the parties or their attorneys with a list of arbitrators from which to select (usually a panel of three). When the selection process is complete, the arbitrators then conduct the arbitration according to established practice and written rules.

In the arbitration process, there are limitations on discovery of the facts, i.e., on the prearbitration disclosure of pertinent facts or documents. Therefore, the owner (assuming it is the adversary party) should seek as much informal discovery from the contractor as possible. Given the fact that the contractor's claim is usually based in large part on its own documents and records, an owner bound to defend the claim in arbitration may be at a disadvantage unless it is allowed access to the contractor's documentation.

Arbitration proceedings are more informal than proceedings in litigation. Rules of evidence do not exist, and the arbitrators are not bound to accept or reject any established rules of law. In spite of these relaxed procedures, the award of an arbitration panel will generally be enforced in a court of competent jurisdiction and will not be overturned for a lack of evidence, unless gross negligence, fraud, or corruption on the part of one or more of the arbitrators is established.

The arbitration proceeding may be similar to litigation. In each proceeding, the parties present their case to the arbitration panel (or judge or jury, in the case of litigation). In each, the parties attempt to persuade the arbitration panel of the correctness of their cause by testimony, documents, and argument. Each proceeding will result in an enforceable judgment.

Arbitration allows parties to present their positions to a single arbitrator or panel of arbitrators while avoiding many of the drawbacks and technicalities of the normal judicial process. However, arbitration also has its drawbacks. The advantages and disadvantages most frequently cited are described in the following paragraphs.

Speed of resolution. Arbitration may avoid the lengthy waiting period common with court calendars. However, arbitration can be a more time-consuming process once it is started. The difficulty of scheduling continuous arbitration sessions because of other demands on the arbitrators' time can result in long delays between sessions, and the process may drag out over many months in a complicated construction case.

Cost savings. Arbitration can be less costly than a comparable court proceeding, but this is not always the case. Construction arbitration almost always involves attorneys. Attorneys presenting a case for arbitration should be as prepared as if the case were tried before a judge or jury. Preparation requires examining documents, interviewing witnesses, performing legal research, preparing experts, and developing demonstrative evidence. All these costs are common to trial preparation. In addition, there are arbitrators' fees of $500 per day or more, administrative fees, and the cost of meeting rooms. These costs are unique to arbitration and generally exceed typical court costs.

Technical expertise of the arbitrators. Presentation of construction claims to a panel of experienced construction arbitrators is generally cited as an advantage of arbitration. Ideally, panels selected for construction disputes consist of a contractor, an engineer, and a construction lawyer. Frequently, however, the parties' attorneys do the selecting and tend to load the panel with other attorneys.

Privacy. Arbitration has the advantage of privacy. Arbitrators are not required to issue written decisions, and decisions are not a matter of public record. The absence of a written decision is of little concern to most since the primary goal of the disputing parties is the prompt resolution of the dispute.

Avoiding legal technicalities. In arbitration, strict rules of evidence do not apply. Arbitrators are generally liberal in their acceptance of evidence. This permits an easier and faster presentation of records, correspondence, documents, photographs and testimony. The AAA rules encourage arbitrators to accept all evidence that may shed light on the dispute. Arbitrators retain the right to determine the weight that they will give to each piece of evidence. They may choose to disregard an item of evidence even after it has been submitted for consideration.

Lack of discovery. Formal discovery procedures are not used in arbitration, unless agreed to by the parties, specifically authorized by statute, or ordered by a court or arbitration panel. Prehearing document exchange is virtually always ordered by the arbitrators. An inability to compel discovery can be either a drawback or an advantage, depending on the status of a party's own knowledge of the facts and documentation.

Location of the arbitration. The location of court proceedings is determined by complex rules of jurisdiction and venue. In arbitration, the parties can use the hearing locale established in their contract.

Waiver of jury rights. Although a panel of technically skilled arbitrators may be a good tribunal for a complex construction dispute, parties to an arbitration agreement should remember that they have waived their right to present their arguments to a jury.

Limited scope of judicial review. Because arbitrators function in a quasi-judicial role, their award will not be overturned by a court unless there is proof of fraud, undue bias, or corruption.

Litigation

Litigation is an expensive and time-consuming process. Much of the lawyers' time spent handling a construction claim deals with procedural issues, such as discovery, motions, and pleadings as discussed later. These procedural issues often drag out the case for a year or more.

The rules governing court proceedings are quite complex and vary among federal, state, and local courts from jurisdiction to jurisdiction. Therefore, the party initiating the litigation should have its attorney present a probable scenario of the course of the litigation, the probable schedule, and cost of the process. The party should also consider staff time that will be involved. An analysis of the "cost-effectiveness" of the litigation process would be in order, much like comparing alternatives for a treatment process.

A decision that must be made concurrently with the selection of parties is the choice of the most desirable court. Typically a jury trial is afforded, but in some jurisdictions and in federal court, a demand for a jury trial must be filed in a timely manner. Parties can waive their right to a jury trial. Which option is best depends on the nature of the case and whether a jury may be more sympathetic to the claim. A jury may be better able to understand a single-event case such as an automobile accident or murder, but a typical jury may not understand as well the complexities of a multifaceted construction claim with many technical sidelights.

A judge can hear all of the evidence and render decisions without a jury. A judge sitting alone is more likely to admit disputed evidence that would normally be excluded from a jury. In such a case, the judge is relying on experience and training to give any evidence the weight it deserves. On the other hand, the judge is not likely to be swayed by sympathy factors, which are frequently played to before a jury.

Discovery. Discovery is the statutory means by which a party is able to elicit facts from the opposing party. The process of discovery is very important and the point at which many construction cases are won or lost. Attorneys must discover every fact, every document, every witness, every photograph, and every other item that will assist in presenting the case. Discovery is designed to eliminate courtroom surprise and encourage settlement. The process can begin as soon as litigation starts.

The discovery process includes the following elements.

Interrogatories. Interrogatories are answers prepared in response to written questions.

Depositions. Depositions are the oral examination of any person with knowledge pertinent to the case. Depositions are taken under oath and recorded by a stenographer. The transcript can be used at the trial. An opportunity to cross-examine is given to the other party.

Admission. Admission entails one party requiring another party to admit, in writing, the truth of certain facts.

Documents. The production rule allows one party to compel another to furnish designated documents or tangible items that have a direct bearing on the case.

Trial. The purpose of the trial is not to demonstrate to the judge or jury how much work has been done in preparation but to present the case in its simplest, most persuasive form. The core of the case involves the resolution of two distinct issues, liability and damages. Has the defendant performed or failed to perform an act or acts for which it was obligated, to the plaintiff's detriment? If so, has there been any resulting damage to the plaintiff and in what amount?

Trial tactics, strategies, and procedures are beyond the scope of this discussion. However, the approach used in a trial is mostly directed by the legal counsel that is retained for the litigation process. The preparation for the trial, the presentations during the trial, and the closing arguments are under the control of the counsel. The counsel's primary goal throughout the entire process is to present the facts that substantiate the position of his or her client.

Costs of Resolution

Trials and arbitration are both costly. The cost of either approach should be estimated beforehand with the construction claims attorney. It is important to know or at least have a good idea as to the possible costs. The party should never hesitate to question its attorney, both before and during the representation, concerning legal fees, as well as other expenses for expert witnesses, documentation, discovery, and other related costs. There are several areas where cost savings can be realized, and these should be pointed out in the beginning and investigated throughout the process.

ALTERNATIVE DISPUTE RESOLUTION

Alternative dispute resolution (ADR) refers to methods of resolving construction-related disputes without resorting to either arbitration or litigation. The interest in ADR in the construction industry is increasing for many reasons, some of which are

- the extremely high cost of either arbitration or litigation
- the need to resolve the dispute more quickly than by either arbitration or litigation
- the clogged court system, which encourages ADR as a fair way to resolve disputes and also relieve the courts
- ADR's ability to balance the desire for speedy resolution against the desire for just resolution

ADR is not an attempt to bar the litigant's right to a jury trial under the Seventh Amendment to the US Constitution. Rather, it is an alternative for complex construction-type disputes that are more appropriate to nonjury trials. Because parties in construction claims cases are less likely to rely on winning their position by playing to a jury that is made up of randomly selected laypeople, these cases lend themselves more to nonjury trials.

Mediation

Mediation is a process that uses an impartial mediator to assist the parties in reaching a settlement by suggesting ways to resolve the dispute. Unlike an arbitrator, a mediator does not render a decision and has no power to force the parties to accept a settlement. A variety of procedures and techniques are used by the mediator to help the parties reach and agree on a resolution.

Mediation has been increasingly used in the last five years or so to resolve disputes in the construction industry, particularly in federal courts. Judges have encouraged mediation as a trial alternative. The mediation process hastens the settlement, reducing the costs to the litigants and the burden on the court system.

The mediator's role can take various forms. The mediator can

- identify and narrow issues
- identify underlying interests and concerns
- carry messages between the parties
- explore bases for agreement and the consequences of not settling
- develop a cooperative, problem-solving attitude in the parties

The advantages of mediation include the following:

- The cost is usually much less.
- Communication between the disputing parties is enhanced through a skillful mediator.
- Personality issues are kept out of the process to a large extent.
- The mediator can work behind the scenes to structure a framework for settlement.
- Mediation can narrow issues in a dispute even if a full settlement is not reached. This can be very helpful in complex cases that include many disputed issues.

Mediation is only as good as the parties' willingness to work at it. Even though parties enter into mediation, they may not be satisfied unless they win their position completely.

The following paragraphs describe the steps that usually occur in mediation.

Agreement to mediate. A trial judge can impose or suggest mediation, often at the same time a trial date is set. The parties can also agree to mediate. They typically sign a written agreement that deals with such matters as identifying the mediator, fees, and confidentiality. Mediation for the construction industry is offered by the AAA, which first published rules in 1984. If the AAA receives a request to mediate, it will appoint a mediator. Several private firms offer arbitration and mediation services specifically to the construction industry.

Mediator selection. Upon receiving a request for mediation, an agency will appoint a qualified mediator to serve. Most likely the mediator will be an attorney. Normally a single mediator will be appointed unless the parties agree otherwise.

The appointed mediator from a mediation agency or the AAA should have expertise and experience in construction litigation, although nonattorney experts are also active in the field. In addition, the mediators should have received a substantial amount of mediation training and should typically have nonjudgmental personalities. The credentials of the proposed mediator should be carefully reviewed. Criteria for selecting the appropriate mediation agency to furnish the mediator and rules of mediation include

- experience of the agency
- training and experience of the mediators
- agency's success rate
- fees and references

Mediation process. Once into mediation, the mediator makes all arrangements with or between the parties. Each side is represented by a person who has adequate authority to enter into a settlement agreement. This means that clients, as well as their attorneys, should attend the mediation session and that a public water agency should be represented by a high-level manager who has the authority from the board to make decisions. The mediator conducts joint and separate meetings with the parties, reviews the positions, reviews supporting data, and may shuttle back and forth between the parties.

If necessary, the mediator may obtain expert advice on technical issues, provided the parties assume the expenses of obtaining this advice. Mediation sessions are private. The mediator cannot divulge confidential information disclosed by the parties. All records and reports are kept confidential. No stenographic record is kept of the mediation process. The expenses of witnesses for either side are paid by the party producing the witness. All other expenses of the mediation, including travel, are borne equally by the parties unless they agree otherwise.

Mediation ends when (1) a settlement agreement is reached by the parties, (2) the mediator issues a written declaration to the effect that further efforts at mediation are no longer worthwhile, or (3) either party issues a written declaration that the mediation process should be terminated.

Summary Jury Trial (Mini-Trials)

A summary jury trial, also called a mini-trial, is a nonbinding, informal settlement process in which jurors hear abbreviated case presentations, typically lasting no more than a day. A judge presides over the hearing. No witnesses are present and the rules of evidence are relaxed. Important to the process is that party representatives authorized to settle the case are *required* to attend. After the mini-trial, the jurors deliberate and hand down an advisory verdict, which becomes the starting point for settlement negotiations. The presiding judge may participate in these negotiations. The verdict is nonbinding upon the parties and is used by the parties to predict the likely outcome of an expensive jury trial.

A summary jury trial is a more elaborate ADR method that requires the attention of a judge and a jury. It is best suited to large, complex cases that would take weeks or months to try, as well as in cases for which the savings in time and money resulting from a successful summary jury trial would be the greatest. A judge cannot prevent parties from getting their day in court but may require them to submit to a summary jury trial before they may proceed to a full jury trial. When such a procedure is imposed, many cases are settled before reaching the actual trial stage.

The real purpose of the summary trial is to get to the essence of the case in as short a time as possible. This forces the parties to evaluate their cases and get to the core of their position. It enables the parties to view how their cases may be seen when examined in the full light of a jury. It also provides each party the opportunity to view the opponent's case.

The advantages and disadvantages of summary jury trials include the following:

- The strengths and weaknesses of each party's position are revealed, which permits a better evaluation of each position.
- Summary jury trials provide a less costly, nonbinding process that encourages settlement. The required involvement of the "decision makers" and principals is important in reaching a settlement.
- The summary jury trial is supposed to be indicative of a full jury trial, but this may not be the case since each jury is unique. Juries are highly unpredictable and often seem to favor large settlements from the party they perceive to be "wealthy."
- Since the summary jury trial reveals the essence of each party's case, there is the need for mutual good faith.
- Despite the results of a summary jury trial, the case may still go to court, in which case the cost of the summary jury trial becomes an additional expense.
- A summary jury trial has limited usefulness and is best suited to large, complex cases that would involve months of court time.

Dispute Review Board

A dispute review board (DRB) generally consists of three members selected and approved by both the contractor and the owner at the time a construction contract is executed. DRB members are experienced with the type of work involved, respected by their peers, and neutral. They visit the job regularly during construction and are kept advised on its progress. When disputes arise, the DRB can recommend settlement quickly, before adversarial attitudes grow and harden. Experience has shown that the existence of an impartial, respected panel encourages parties to view their differences more objectively and to resolve them without a hearing. The DRB provides an independent assessment on the merits of disputes and does not usurp the owner's authority to direct the work as provided in the contract. Although DRB recommendations are not binding, parties normally give them great weight and in most cases accept the findings. Should further proceedings such as arbitration or litigation become necessary, those recommendations are admissible as evidence.

The DRB is not intended to supplant existing dispute settlement methods. Rather, it is an earlier, nonbinding, intermediate step directed at avoiding the need to resort to more expensive, more time-consuming, and less satisfactory procedures. On June 1, 1993, a new set of Construction Industry Dispute Review Board Procedures, administered by the AAA became effective.

Chapter 14

Payment Procedures

The term *payment procedures* refers to the process of a project owner paying a contractor for work performed to date. Payments signify completion of certain portions of a project and imply acceptance of the work performed to that point.

The contract should define the specifics of the payment procedure that will be followed. The procedure should be presented in sufficient detail to eliminate ambiguities and misunderstandings by all participants in the project. In practice, the contractor, the consultant, and the owner often establish a payment schedule after the contract is awarded. However, specifying the basis of payment in bidding documents benefits all participants. The payment procedure is determined by the type of project, funding sources, length of project, complexity of the project, and the owner's existing framework for making payments.

Because progress payments are made as the work progresses, the quality of work as a function of the completed project cannot always be fully evaluated. In many cases overall project quality cannot be determined until the project is completed, started up, and tested. Payment methods and procedures can affect the risk assumed by the contractor in the progress of the work. Reducing the risk to the contractor will aid the contractor in its bid preparation and thereby reduce costs for unknown complications.

This chapter examines these issues in more detail. Included in the discussion are

- payment methods
- payment schedule
- payment procedure work flow
- forms
- basis of payment
- retainage
- completion
- project startup
- late payments

PAYMENT METHODS

The most common payment methods are reviewed in the following paragraphs. Each method has different payment procedures that will be discussed in this chapter.

Lump Sum

In a lump sum contract, the contractor agrees to perform work for a set price. Such a contract may also be referred to as a fixed-price or stipulated-sum contract. This type of contract gives the owner control over the total cost of construction and shifts more of the risks of construction to the contractor. Lump sum contracts are more applicable to smaller projects, those completed in a few weeks or with only one or two components.

Both owner and contractor face potential problems when using a lump sum contract. A precise scope of the work is required. The need for complete drawings and specifications might increase the lead time from the start of design to the start of construction.

With this type of contract the contractor assumes the risk of being able to perform all of the work for the amount specified. Because of this type of risk, a contractor may include additional amounts or contingencies in the bid to compensate for any increased costs due to unknown conditions.

If a lump sum project takes several months to complete, the contract usually allows for progress payments based on percentage of the work completed.

Unit Price

The unit price payment procedure requires the consultant to provide a quantity takeoff—an itemized list of bid items—as part of the contract documents. This quantity takeoff provides the contractor with the first step toward developing a bid. Additionally, unit prices assist in processing change orders should additions or deletions be made to the project during the construction period.

Under this approach, the risk to the contractor is reduced somewhat relative to a lump sum payment. The project cost to the owner may vary if the estimated unit quantities of work differ from the work actually performed by the contractor.

Cost Plus Fixed Fee

Cost-plus-fixed-fee contracts provide for (1) reimbursing the contractor for direct and indirect costs incurred in the project construction and (2) paying a fixed fee. The additional fee serves as the contractor's profit and is negotiated when the parties enter into the contract.

Municipal competitive bidding requirements usually restrict this method to privately owned construction. In construction of privately owned facilities, the owner might select the contractor because of good past performance. This allows the owner and the consultant to negotiate a fixed fee the contractor will receive in addition to construction costs.

This method might result in a lower project cost because the contractor's risk now becomes a *cost* to be reimbursed instead of an estimated cost that is part of a bid. The contractor's risk is eliminated from the bidding process unless the project extends beyond the estimated total cost or completion time. However, the owner's cost will vary as the actual construction cost varies from the consultant's estimate.

Time and Material

Time-and-material contracts provide for (1) a fixed hourly rate for the number of hours the contractor spends performing the contract and (2) reimbursement for the contractor's material costs. The owner negotiates the labor rates to be applied to the construction work involved in completing the project. The contractor is reimbursed for cost of materials at a fixed markup and is paid for the labor at agreed rates. Privately owned utilities use this method more than public utilities.

This method can provide the lowest project cost because the risk to the contractor is eliminated, since the labor and material costs plus profit are assured. However, there can also be a lack of incentive for the contractor to limit the construction costs. For this reason, labor and material contracts are often used only when no other type of contract is applicable. For example, municipalities often maintain time-and-material contracts for emergency repairs that must be initiated without plans or specifications being available.

Combination of Methods

Depending on the type of project, a combination of (1) the unit price and cost-plus-fixed-fee methods of compensation or (2) the lump sum and unit price methods is often advantageous. For example, if unknown foundation problems occur in a concrete tank construction project, it might be advisable to use unit prices for excavation and concrete work, while using a lump sum for the instrumentation package. If more work than anticipated is required, a unit price exists for that extra work; and if less work is involved, the owner saves some money, since only the actual work required is reimbursed. The contractor is also assured of receiving payment for the actual work involved.

PAYMENT SCHEDULE

The payment schedule—that is, how often the contractor submits a payment request—should be specified in the contract. Most payment schedules are based on a calendar month. It is possible to process payments in 60- or 90-day cycles, but the contractor might have to include interest costs into the project bid for the added time it must finance the work before reimbursement.

If outside funding is scheduled to be received from federal or state agencies, the owner should fully understand the scheduling of how these funds are to be dispersed. Questions to be resolved include:

- Is the funding based on a reimbursement of the owner's costs after the contractor is paid by the owner? If so, what is the schedule for reimbursement—say, 30 days after the request is received from the owner, or possibly on a quarterly basis?
- Is the funding to be received as a fixed amount on a fixed schedule?
- Are the contractor's payment requests to be submitted for review and approval by the funding agency before action by the owner?

When establishing the payment schedule, the owner should review its accounting procedures. For example, on what days are checks prepared by the owner's accounting department? Do checks then need to be submitted to a city council or board of directors for approval before payment?

Other contractual items to be specified include

- the exact day of the month for the contractor to submit the payment request
- to whom the payment request is given (for example, the construction management agency, the owner's project representative)
- the number of copies of the payment request
- the time for review and approval or revision of the contractor's payment request

The contractor and the owner's field representative may jointly develop quantities in the payment request, but the contractor should complete and sign the payment request form. Depending on the complexity of the project, the consultant's review time may increase as the project progresses. It is prudent to allow sufficient time in the work flow for complex payment request submittals to be reviewed. Often the city council or board members want all actions, such as a payment request, submitted at least five days before their meeting. If each monthly payment requires council or board action, the work flow should allow sufficient time.

After payment procedures are developed and payment-processing time established, contractors have the right to expect that those procedures will be followed and that they may anticipate payment within the agreed period.

Figures 14-1 and 14-2 illustrate some sample payment procedure work flows.

IMPORTANT ELEMENTS

This section describes the important elements in project payments.

Payment Request Form

To simplify review and processing of the payment requests, either the owner or consultant should develop a payment form for the contractor to use. The form is included in the contract documents. The owner might use the same payment form on all contracts to simplify review and processing.

Payment procedures used in standard contract documents should be evaluated to make sure they fit the project as well as the owner's operating procedures. The owner should be aware that its own existing payment procedure may not fit the "standard" payment procedures of others.

Basis of Payment

To avoid conflict or confusion, the basis of payment to the contractor should be established. The main types of the basis of payment are described next.

Completed work. An established unit price may be applied to estimated units of completed work such as cubic yards or lineal feet. Alternatively, a lump sum, such as for a piece of equipment or contractor move-in, or a percentage of a lump sum contract may be the basis of payment.

Materials on-site. If payment will be made for materials that are on-site but are not yet incorporated in the project, contract documents should define how and where materials are to be stored. Documentation that must accompany the payment request for material stored on-site, such as delivery receipts or invoices, should be specified.

It is important for the contract to state that the safety of stored materials, even if paid for by the owner, is the responsibility of the contractor and that if materials are damaged on-site, the contractor is responsible for replacement. The contract

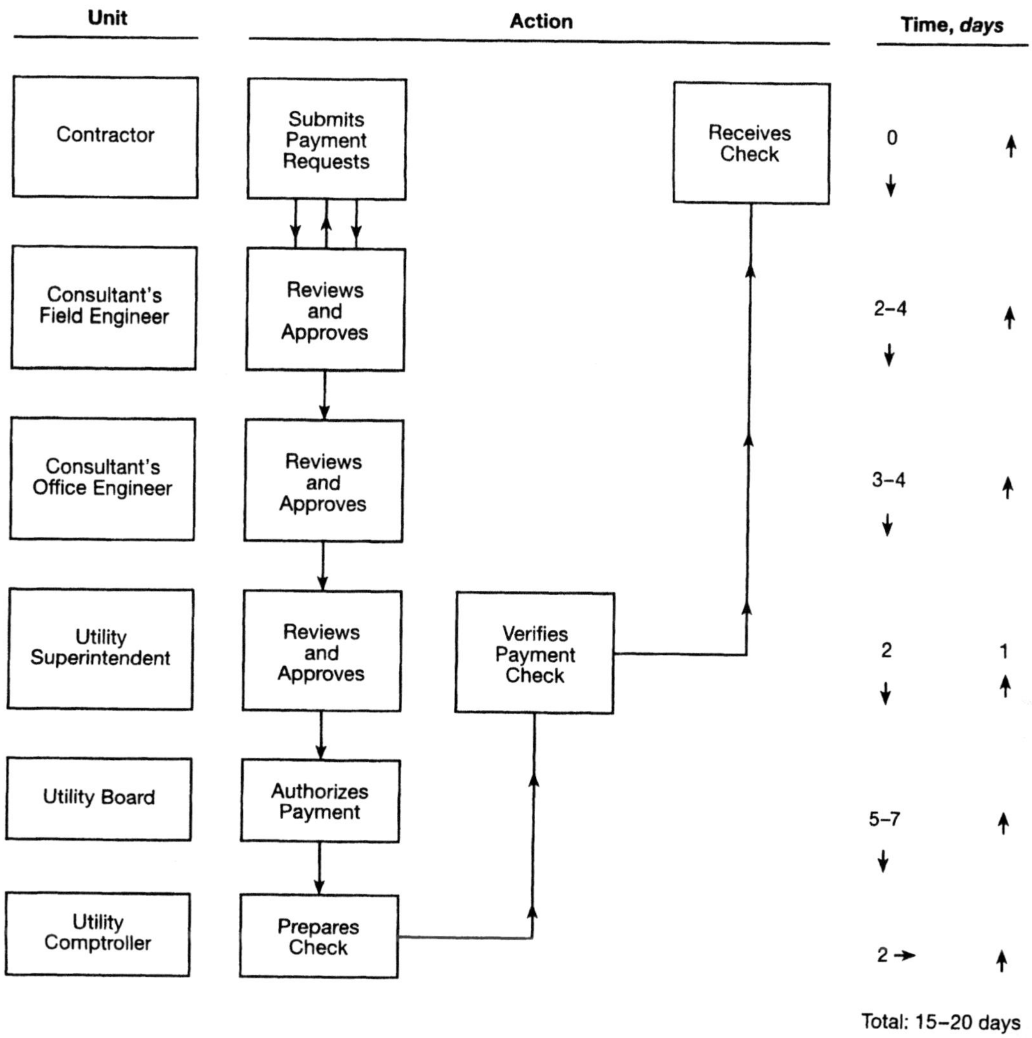

Figure 14-1 Sample payment procedure work flow

should also stipulate that the warranty or guaranty dates commence on final acceptance and not on delivery to the site.

Required Documentation

Contract documents should clearly specify what documentation shall accompany the interim and final requests for payment, such as paid invoices, delivery receipts, and release of liens from subcontractors. The owner might need to require an assurance that prevailing wages are being paid.

In states that have lien laws, the contractor must also submit to the owner a statement that gives the names of all those who are furnishing materials or labor for the project, the cost of these materials and labor, and the amounts due or to become due each of them.

Documents should specify that the owner shall receive all manuals, spare parts, and startup training on specific equipment items before final payment is processed.

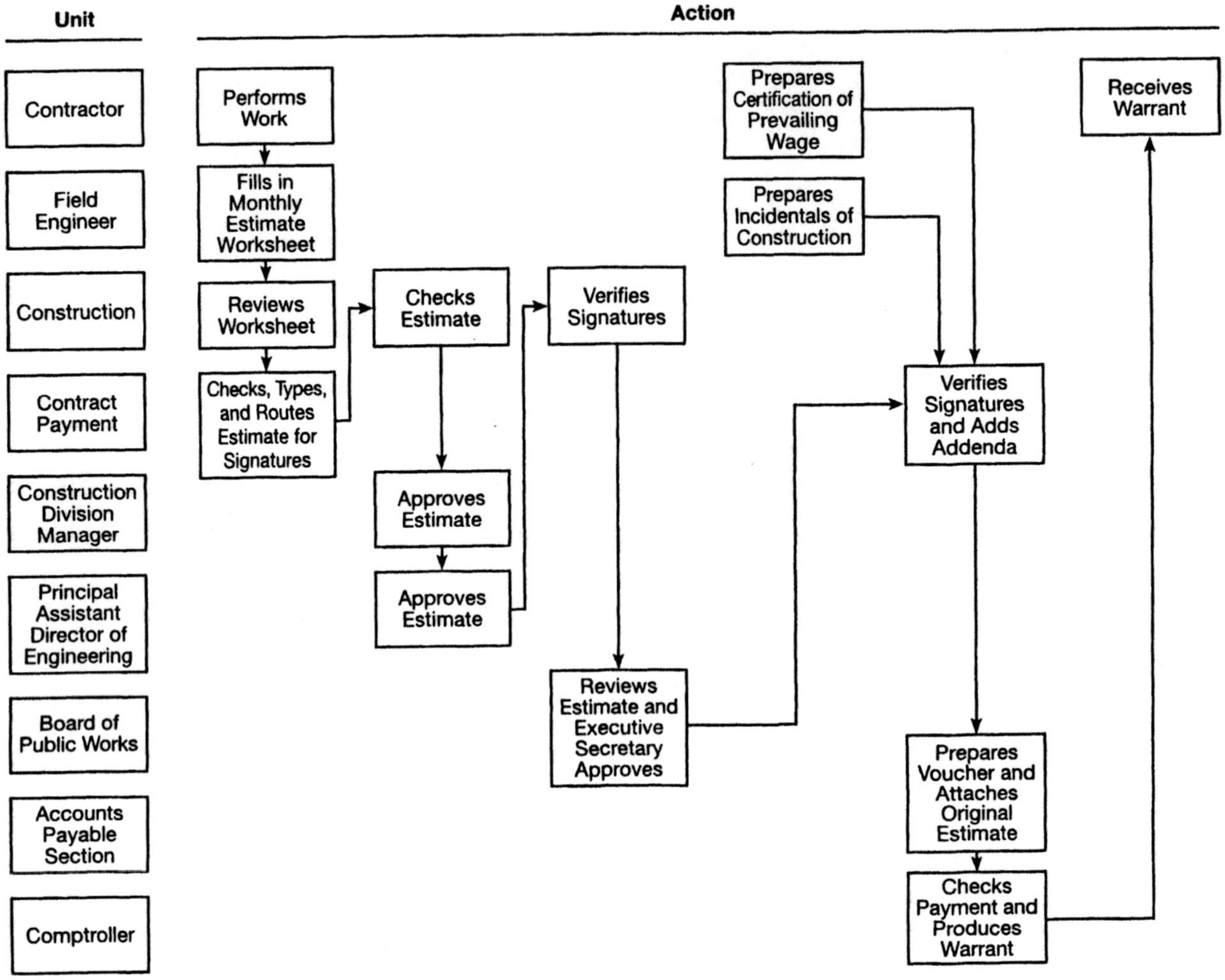

Figure 14-2 Sample contractor payment work flow

Retainage

Contract documents must specify the retainage (chapter 5 also discusses retainage) to be held from each progress payment. Retainage is usually a percentage withheld from the payment request as a guarantee of the quality of completed work.

Statutes of many states permit the contractor to specify how the owner shall invest the retainage and that the contractor is entitled to the interest accrued on the retainage. The owner and the contractor must understand and agree on the procedure for holding and releasing the retainage. Conditions for releasing the retainage are discussed later in this chapter.

PAYMENT PROBLEMS

If the procedures discussed so far are followed, then the potential for problems is reduced. However, a few areas for problems or disputes are discussed next.

Overestimating Progress

Overestimating the value of the work completed by the contractor (also known as front-loading) may cause budget deficits in the final phases of the project, particularly with cost-plus contracts. Occasionally, the contractor overloads the value of the initial construction work to cover the costs of project move-in and to provide up-front funding for the contractor's project costs. The contractor is entitled to true move-in costs and true reimbursement for work completed but not excessive front-loaded funding. Often a fixed dollar amount available to the contractor for the first payment request is given in the bidding specifications to let the contractors making bids know what the move-in funds will be.

Underestimating Cash Flow and Late Payments

The owner should have the funding in place to match the contractor's estimated progress payments. To this end, contract documents might require that the successful contractor submit to the owner's field representative an itemized work schedule for complete construction of the project, as well as an estimated monthly payment request, along with the initial payment request.

The owner should advise the contractor in writing immediately if a payment will be delayed or reduced in amount and it should state the reasons for the discrepancy. If the payment delay is due to the owner's problems, such as funding reimbursement delays, the contractor may be entitled to interest on the delayed amount.

If the payment delay is due to the contractor's operations, the owner should advise the contractor of the problem and the actions required by the contractor to remedy the problem. All payment delays or deductions in processing payment requests should be documented in writing to all parties involved.

PROJECT COMPLETION

In the final phases of a project, generally referred to as completion, payment procedures are different from progress payments. Various terms for completion are defined below.

- *Substantial completion:* that time in the project when the owner has full and unrestricted use and benefit of the facilities, from both the operational and safety standpoint, and only minor incidental work, replacement, or correction work remains to reach physical completion of the project.
- *Final completion:* that time in the project when all physical work is completed and accepted; and all operational testing has been satisfactorily completed. The final payment, less retainage, may be made at this time.
- *Final acceptance:* that time in the project when all physical work and operational testing have been satisfactorily completed and the contractor has furnished all documentation required by the contract and by federal, state, or local regulations. The release of the retainage may be processed at this time.

Even though the consultant, owner, and contractor have discussed payment procedure throughout project construction, the completion of a project, processing of final payment, and release of retainage are often a source of conflict.

Steps in a typical project completion are as follows:

1. The contractor submits a notice of substantial completion to the owner.
2. The consultant and owner review the work and prepare a list of items yet to be completed.

3. Operational testing of facilities, if required, is scheduled.
4. The contractor completes all physical work, and all operational testing is *satisfactorily completed. Startup services and personnel training are completed, if required.*
5. The owner processes final payment less retainage.
6. The contractor submits all documentation and spare parts required by the contract, such as warranties, release of lien forms by subcontractors and suppliers, and certification that prevailing wages have been paid where required.
7. The owner authorizes release of retainage.

Chapter 15

Private-Sector Financing

Although project financing is not necessarily a direct part of contract administration, without financing there would be no project to construct. The demand for capital improvements to water systems and the shortfall in state and federal funds has created a need to expand funding sources.

Private sector financing provides an alternative to reliance on conventional sources such as grant programs and municipal bond offerings. This chapter identifies circumstances under which private-sector financing may provide a favorable alternative to traditional public-sector financing. Private-sector financing may take several forms, including privatization, lease purchase, or developer contributions.

PRIVATIZATION

In its broadest sense, privatization refers to private sector service on behalf of a public agency. Privatization has come to mean a range of activities from the financing, designing, construction, and operation (the last three of which make up what is called a turnkey project) of a project to specific contract services provided by a private enterprise for a municipal system. Under privatization, compensation is linked to performance rather than a fee for simply providing the service. Other chapters discuss design and construction aspects of privatization.

For many years the private sector has provided funds for main extensions through developer contributions and system development charges. However, use of these funds is generally limited to only the portion of improvements necessitated by growth. Improvements due to old age, obsolete technology, regulation, and general renewal and replacement are not funded by system development charges.

Developer contributions and traditional public-sector financing methods alone are insufficient to meet the growing need for capital improvements. Although there remains a strong demand for traditional public-sector financing, there are a growing number of situations in which private-sector financing may be preferable as the best overall solution. This chapter identifies those situations and explains the relative merits of public versus private financing.

Even in the low-interest municipal borrowing market that existed in 1995, private-sector financing had a role in funding public projects. Historically, governmental

Table 15-1 20-year maturities—borrowing cost

	20-year maturities—borrowing cost		
Rating	Aaa	Aa	A
Tax-exempt	5.30%	5.40%	5.60%
Taxable	7.20%	7.30%	7.45%

agencies have been able to finance water plant construction at a lower cost than the private sector. This has primarily been the result of the significant interest rate advantage held by governments because of their ability to issue tax-exempt debt. In December 1993, for example, a government could borrow money at 74 percent of the cost that would apply for a comparably rated private firm. Table 15-1 compares December 1993 government tax exempt rates with private taxable rates for the various bond ratings. In addition, governments frequently finance 100 percent of the cost of most facilities, whereas private utilities are expected to invest some equity, thereby limiting the amount that private utilities can finance with lower-cost debt funds. An equity portion of at least 20 percent of the total project cost is generally a requirement imposed by lending agencies.

During the period 1979 to 1986, the combination of investment tax credits and accelerated depreciation made it possible for private financing and ownership to be less costly than government financing. However, the 1986 Tax Reform Act removed these advantages, once again making private financing more costly.

Private ownership (in part or in whole) can shift the borrowing burden to the privatizer, enhance the balance sheet of the public agency, and ease the pressure on the borrowing or debt capacity of the public agency. In addition, overall capital borrowing requirements associated with a specific project may be lower because of the ability of the privatizer to reduce the time and cost associated with design and construction activities.

Even though the private sector may in some cases provide service directly to water users, the responsibility and mandate for service remains with the government agency. Privatization (in one or more forms) is just one technique available to public sector management; it may or may not be the best option in a particular situation.

The success of a privatization project requires a clear understanding of a number of issues by public officials responsible for managing the process. These issues include personnel policies for treatment of displaced employees, necessary training to provide government managers with new skills, including accounting systems to measure performance, and an understanding of private-sector motivation. Benefits of privatization include

- potentially greater efficiency in design, construction, and operation
- reduced time frame to bring a project on-line
- minimized administrative procedures required under federal and state grant or loan programs
- timely and potentially lower-cost capital
- ability to have local control of planning and design without restrictions imposed under federal and state grant or loan programs
- potential for lower cost of operations and consequently lower rates
- no need to develop expertise and operating capability where new major facilities or technology is to be brought on-line

Barriers to privatization include

- complexity of developing and implementing agreements that protect the interests of both parties

- difficulty with procurement restrictions that lean toward selection of lowest bid with limited consideration of bidder's qualifications
- concerns over terminating long-term agreements if service deteriorates
- resistance to turning over responsibility of municipal service operations to a private entity
- concern over losing control of utility operations

STRATEGIES TO ATTRACT PRIVATE-SECTOR FINANCING

A reasonable assurance of adequate profit must exist to attract private-sector capital for financing and construction of production, purification, pumping, and transmission facilities that benefit the system as a whole. Several strategies are available. For example, the Tax Reform Act of 1986 placed restrictions on the tax-exempt status of certain improvements to public-purpose facilities. Advocating tax incentives to allow the use of tax-exempt bonds would reverse this trend and generate additional funding.

The Economic Recovery Tax Act of 1981 stimulated interest in privatization by providing tax incentives for capital investment to private investors. Subsequent tax law amendments, however, removed incentives for privatization and leasing activities and have limited private sector financing of municipal water facilities. Increasing tax-based capital investment incentives could expand private investment.

The focus of more recent private financing alternatives has shifted from reliance on tax incentives to economic efficiencies, references to the technical and business expertise offered by private firms, efforts to minimize institutional barriers, and modification of regulatory policies.

PUBLIC-SECTOR FINANCING PROCESS

To be able to identify situations in which private-sector financing may be advantageous to a government agency, the tax-exempt financing process and key terms must be understood. The following paragraphs explain the process and define the terms.

Typically, tax-exempt bonds are classified by the revenue stream that is pledged to repay or retire the debt. When bonds are secured by the general taxing power of the government agency, they are referred to as general obligation (G.O.) bonds. When bonds are secured by the revenue stream of the utility (i.e., user charges and other utility revenues), they are referred to as revenue bonds. The market views a G.O. pledge as superior to a revenue pledge and will rate a G.O. bond at one letter grade better than a revenue bond by the same agency. Generally, the better the rating, the lower the cost of borrowing. The reason for this valuation by the marketplace is that there are many more scenarios that could result in inadequate revenues to retire the revenue bond than there are situations in which the G.O. pledge would prove inadequate.

The financing capacity of G.O. debt that can be issued by an agency is typically stated as a percentage of the assessed valuation of the land on which property taxes can be levied. In the case of revenue bonds, the rules are more complex. Issuers of revenue bonds must meet "coverage tests"; i.e., the revenue remaining after operations and maintenance costs are paid must be some multiple of the annual debt service payment, such as 1.10 or 1.25. (The word *coverage* comes from the number of times that debt service is covered by net revenues. A multiple of 1.25 is usually referred to as 25 percent coverage.) It is not unusual to find that a bond of a given size cannot be issued because the coverage test demanded by the rating agencies will result in rates that are deemed to be excessive.

The creditworthiness of the issuer can have a significant effect on the interest rate that must be offered to make a bond salable. As illustrated in Table 15-1, a one-letter grade reduction in rating requires the issuer to offer between 0.1 percent and 0.2 percent additional interest. In cases where an issuing agency is expected to be poorly rated, it is possible to improve the rating through a credit enhancement mechanism. The most common form of credit enhancement is through the purchase of bond insurance. Bonds that are insured by these bond rating agencies will be given a Aaa rating (which essentially corresponds to the rating of the insurer who guarantees to make debt service payments in the event the issuer is unable to do so). However, the insurance premium can be sizable and is usually due up front. Other forms of credit enhancement can take the form of payment guarantees such as backup pledges, developer guarantees, or letters of credit.

Issuing a tax-exempt bond can be a lengthy and costly process. Typically 3 to 6 months or more will be required. Some of the costs and technical expertise involved in bringing a bond issue to market include

- *Financial advisor*—to oversee the process, structure the transaction, help in selecting an underwriter and trustee, and so on.
- *Bond counsel*—to ensure that the process conforms to legal requirements, prepare necessary legal and tax forms, verify that the bonds are tax-exempt, and so on.
- *Feasibility consultant*—to conduct a study to verify that the bonds are likely to be repaid and that the issuer can meet the coverage test. Feasibility consultants will also calculate the water utility rates required to support the bond issue. Feasibility consultants are generally required only for revenue bond issues.
- *Underwriters discount*—the difference between the price at which the underwriter purchases the bonds from the issuer and the price at which the bonds are sold. The discount is how the underwriter is compensated for the cost of marketing the bonds and for the risk assumed by the underwriting team.
- *Engineer*—Bond documents typically include a consulting engineer's report that identifies the need for the anticipated capital projects and support for the estimated cost.
- *Accountant*—Bond documents typically include a full or condensed set of audited financial statements for the issuing agency.
- *Printing and mailing costs*—A large number of bond documents (a document that describes the issue and includes the consultant reports is referred to as the official statement or OS) must be printed and mailed to the bond marketers so that the documents can be provided to potential buyers. Since 1986, the amount of issuance costs that can be included in a bond issue has been capped at 2 percent of the project costs.

Not included in the previous list are the personnel resources (e.g., city attorney, finance director) of the issuing government. These personnel can be significant in a tax-exempt issuance and should not be overlooked. Based on the previous discussion, it is apparent that significant amounts of time and resources are necessary to issue tax-exempt debt. Much of this process can be reduced or eliminated under private financing—one of the tangible benefits of involving the private sector.

The private sector has a more diverse array of financing sources available to it that can, in some cases, reduce the cost and time typically associated with public-sector borrowing. These sources include

- accumulated capital from operations
- sale of equity, in the form of either (1) common or preferred stock or (2) an equity interest or position in specific projects through partnerships or similar arrangements
- borrowing from banks or underwriters; most large firms participating in the "privatization" arena have revolving credit arrangements.
- project-financing vehicles; these are similar to the public sector's revenue bonds in that the purchasers of the debt evaluate both the project and the bond issuer.

However, 100 percent project financing (the norm for public-sector offerings) is rare; the private sector is expected to invest equity, typically around 20 percent of the project, in this type of transaction. Some major distinctions between public and private financing are that private financing generally requires a much shorter transaction time and transaction costs are generally lower.

In comparing alternative financing approaches (e.g., public versus private), analysts generally look at

- *Financing terms*—which primarily consist of maturity period (how many years to retire the bonds), interest rate, payment terms (i.e., level debt service —principal and interest—payments or other), and whether or not there is a capitalized interest period (a period at the beginning of the bond issue when no debt service is paid)
- *Transaction costs*—the costs of an issuance (described previously) plus bond insurance or other credit enhancement costs
- *Transaction time*—the length of time required from the beginning of the financing process until the money is obtained

SITUATIONS SUITED TO PRIVATE-SECTOR FINANCING

Several opportunities or situations lend themselves to private-sector financing. They are discussed next.

Insufficient Debt Capacity

If the government entity has insufficient debt capacity, private-sector financing may be necessary. Though this may appear to be an inescapable conclusion, the facts are often more complicated. In the case of G.O. bonds, it is usually possible for a government entity to obtain waivers to exceed the indicated G.O. debt capacity. However, unless the period during which debt capacity will be exceeded is relatively short, operating under such a waiver can become extremely burdensome.

Similarly, if there is a problem with meeting revenue bond coverage requirements for a short period of time, some remedies are available, such as the capitalization of interest. However, if the coverage problem is expected to persist, alternatives should be sought. Interest capitalization allows the issuer to defer interest payments for a fixed period of time. The "cost" of doing so is capitalized into the total issue amount, increasing annual debt service payments.

In situations of inadequate debt capacity, private-sector financing generally does not yield lower borrowing costs. It merely makes financing possible. Private-sector financing makes the most sense in these situations if it is needed to permit development to occur that significantly improves the economic outlook for the government through an increased tax base or other means.

Limited Transaction Time

Under circumstances that do not allow sufficient time to go through the tax-exempt bond process, privatization with a firm that has a pool of equity funds available and/or arranging for a design–build or build–operate–transfer (BOT) contract may be the only way to meet tight time constraints. Reduced transaction and construction time may partially offset the interest rate differential between taxable and tax-exempt financing. Due to inflation, the long-term trend for construction costs is steadily upward. The ability to initiate a project one year earlier under some form of private sector participation can result in a capital cost reduction of 2 to 4 percent.

Short-term or "bridge" financing mechanisms that may permit a public entity to mitigate the public-sector financing limitations were mentioned in the previous section. Specific categories of these mechanisms may include bond anticipation notes, tax increment financing, revenue anticipation notes, and so on. For tax exempt financings, the borrower is required to anticipate the expenditure of at least 85 percent of bond proceeds within 3 years from the date of issuance.

Other Regulatory Considerations

Some states have regulations that make construction under public ownership more expensive than construction under private ownership as a result of either procurement requirements, contracting limitations, prevailing wage rate requirements, or other limitations. The following is an example of how a construction cost "penalty" can more than make up for the interest rate advantage enjoyed by governments.

Consider the following assumptions:

- plant cost of $10 million
- 100 percent financing
- 20-year term, single annual payment debt service
- comparable credits, i.e., Aaa or Aa, for both taxable and tax-exempt status

The annual payment on $10 million for the tax-exempt Aaa credit at 5.3 percent interest would be $822,963. The annual payment for the taxable Aaa credit at 7.2 percent interest would be $958,652, or approximately 17 percent more based on a direct comparison of principal and interest for each case. For the payments to be identical the plant would have to be constructed for slightly less than $8.6 million or about 14 percent less. In some states, however, public construction may be more than 14 percent more expensive than private construction.

The foregoing comparison is much more simplistic than a situation faced by a decision maker in the public sector. In order to arrive at a more realistic comparison, it is necessary to include other cost factors frequently encountered in financing. Assuming again 100 percent financing and a 20-year term with a single annual payment, the public entity is likely to face the following additional costs:

- 2 percent issuance cost
- 1 percent cost of credit enhancement or rating fees
- 2 percent transaction time penalty
- 1 percent administrative costs (additional insurance costs and post-issuance administrative costs)

These costs effectively increase the $10 million facility to $10.6 million. At 5.3 percent interest, the annual payment is $872,341.

Assume the private owner can achieve the following:

- interest earnings during construction 1 percentage point in excess of the amount permitted the tax-exempt owner
- construction cost reduction of 7 percent due to shorter construction period and improved construction management.

At an interest rate of 7.2 percent, the annual payment would be $881,960, or essentially the same amount as under the tax-exempt issue.

The foregoing comparison neglects tax effects as well as the private-sector owner's ability to shelter income through the recognition of depreciation expense. In some states the total marginal tax rate (federal, state, and local) can exceed 50 percent so that a sharing of depreciation benefits could permit private financing in this example to result in a lower annual cost to rate payers. (NOTE: The previous statement is correct only if the private owner is not subject to rate regulation. This additional complexity will be examined in the following section.)

Weak Government Credit, Strong Private Credit

The spread between private and tax-exempt ownership narrows significantly if a poor tax-exempt credit is compared to a strong private issuer. For example, for all of the same factors identified in the previous section but with the tax-exempt entity changed to a B-rated issuer paying 5.8 percent interest, the annual payment for the tax-exempt owner becomes $909,208, or some 3 percent more than under private ownership.

Up to this point the profit necessary to attract private-sector ownership or financing has not been considered. In some cases a true privatizer (one who owns and operates a facility) may be satisfied to make its profit from the operations contract. However, if the private entity will serve only as the owner of the facility and has no other interest in the project, the owner will appropriately require that some profit be realized on the transaction. Furthermore, if this occurs in a state in which the private entity will be regulated by a public service or utility commission, the owner will be restricted to a return authorized by the commission.

To compare the economics of public versus private ownership, it is necessary to make certain assumptions regarding the $10 million plant.

For governmental ownership, assume

- 100 percent financing
- issuance and related costs totaling 6 percent
- 5.8 percent interest rate (B-rated issuer)

These assumptions yield an annual payment of $909,209.

For private ownership, assume

- 20 percent equity investment requiring a 15 percent return
- remaining 80 percent borrowed at 7.2 percent
- interest benefits and construction cost reduction equals 8 percent
- straight-line depreciation allowed over 40 years

In this highly simplified example, calculated on a utility rate-making basis, the annual costs of interest, depreciation, and return under private ownership are $1,035,920, or nearly 15 percent greater than under government ownership. This occurs primarily as a result of the return-on-rate base requirement. Some state utility commissions use the concept of an imputed debt-to-equity ratio or capital structure that assumes the owner had actually made the imputed equity investment, generally in the 40 percent range. In those states the difference in the initial annual payments would be even greater than indicated in this example.

When profit or regulatory rate-setting considerations are introduced, it becomes necessary for the private-sector owner to achieve construction cost reductions slightly in excess of 15 percent to match the annual payment of a tax-exempt owner.

Other Factors

A number of other factors have resulted in private-sector financing proving to be more desirable than public-sector financing. These factors tend to be dependent on the situation, and so the only guideline that can be provided to the decision maker is to remain aware of the possibility that these factors may be present. They include the following circumstances:

- The prospective owner is already constructing a project nearby, and certain economies can be realized through sharing equipment and materials.
- The prospective owner owns land that would be ideal for locating the facility.
- The prospective owner has a planned development that must wait for the construction of certain water facilities. Continued delay will be costlier than a reduced profit on the construction of the facilities through the private sector.
- The prospective owner is interested in entering the contract operations market and is willing to provide financing at a nominal profit to obtain the operations contract for the facilities. (NOTE: Caution should be exercised to be sure the prospective owner has the requisite experience to operate the facilities in question.)

LEASING

Leasing offers the public sector a means of acquiring equipment and facilities when debt limitations or other factors restrict direct purchase and ownership. In addition, acquisition through leasing may allow facilities to be brought on-line with less delay. Leasing, as with privatization, provides the government entity with an additional technique to conserve capital funding projects that may be advantageous in certain circumstances and in some cases may be funded through tax-exempt debt.

Lease transactions can be less complex (i.e., have a shorter transaction time) than traditional public offerings and privatization alternatives; they may afford the municipality greater control over facility operations and maintenance. These benefits are offset by (1) reductions in savings that are potentially available under privatization and (2) the possibility of changes in tax laws and interpretations of those laws by the Internal Revenue Service. Knowledge of tax benefits to the lessor and tax implications to the lessee is essential. Typically, tax-exempt leases are written to allow the government entity the option to purchase the facilities at the end of the contract period for a nominal amount.

There are many forms of lease transactions. Three of the more commonly used forms are sale leaseback, operating lease, and tax-exempt lease purchase. A sale leaseback is essentially a long-term rental agreement that allows for a purchase at various points in the lease period. In a sale leaseback, the investor-lessor provides the capital to construct or purchase the facility or equipment and the lessee operates the facility. An operating lease is similar, but at the end of the lease term the lessee has the option to renew the lease, purchase the equipment or facility at fair market value, or return the property. A tax-exempt lease purchase (TELP) is an installment sale of a facility that has more than 90 percent public-purpose use and is financed at tax-exempt rates. In a TELP, the private-sector lessor obtains no tax benefits from the transaction (in exchange for obtaining tax-exempt financing). In larger transactions, TELPs are sometimes financed through the issuance of certificates of participation

(COP), which provide for payments under this tax-exempt lease purchase to be assigned to a trustee. The payments are then shared among investors proportionally. COP transactions require less transaction time than conventional tax-exempt financing. Because they are sold into a different market, they can obtain more favorable interest rates.

Bibliography

Abbett, R.W. 1963. *Engineering Contracts and Specifications, 4th ed.* New York: John Wiley & Sons.

American Society of Civil Engineers (ASCE). 1996. *Consulting Engineering—A Guide for the Engagement of Engineering Services.* Manuals and Reports on Engineering Practice No. 45. New York: ASCE.

———. 1990. *Quality in the Constructed Project: a Guide for Owners, Designers, and Constructors.* Manuals and Reports on Engineering Practice No. 73. New York: ASCE.

American Water Works Association. 1988. *M29, Water Utility Capital Financing.* Denver, Colo.: AWWA.

Bockrath, J.T. 1986. *Dunham and Young's Contracts, Specifications, and Law for Engineers, 4th ed.* New York: McGraw-Hill.

Bramble, B., B. M. D'Onofrio, and J.B. Stetson. 1990. *Avoiding and Resolving Construction Claims.* Kingston, Mass.: R.S. Means Company, Inc.

Chew, W.H. 1991. A Credit View of Private Financing Alternatives for Water and Wastewater Facilities. *Environmental Finance,* 1(2):251–257.

Corgan, B., H.J. Bell, and W.R. Poplin. 1992. Common Sense Construction Law, Chapter 13, *Resolution of Construction Disputes.* Atlanta, Ga.: Smith, Currie, and Hancock, Attorneys at Law.

Darrow, A. 1983. The Personal, Political and Psychological Aspects of Privatization. Presentation given at *Engineering News Record* Seminars, New York, September 15.

David, I.T. 1991. Apogee Research, *The Privatization Review,* 6(4):17–33.

Engineers Joint Contract Documents Committee (EJCDC) Documents, forms, commentaries, and other publications. Washington, D.C.: American Consulting Engineers Council; New York: American Society of Civil Engineers; Alexandria, Va.: National Society of Professional Engineers.

Flynn, W.T. 1991. Financing Growth and Public Works for the Greater Good, Presentation given at American Society of Civil Engineers (ASCE) specialty conference *Successful Land Development,* September 9–11.

Goldbloom, J. 1992. Improving Specifications. *Civil Engineering,* 62(9):68–70.

Gurry, W.W. 1995. Documenting Design–Build. *Civil Engineering,* 65(9):47–49.

Harley, K.F. 1989. *Privatization V: The Local Privatization Movement in America.* Harrisburg, Pa.: The Commonwealth Foundation for Public Policy Alternatives.

Henderson, K.W. 1992. Use of EJCDC Documents Reduces Conflicts. *Jour. AWWA* 84(9):73–76.

Medioli, A. 1991. Privatization and Municipal Credit: Cautions for Private Entities. *Environmental Finance* 1(3):317–326.

Pfund, N.E. and S.D. Guyer. 1991. Environmental Venture Capital: What Investors Consider. *Environmental Finance* 1(3):291–298.

Shimoda, T.A. (1994). *"... excellent communication skills required" for Engineering Managers.* New York: American Society of Civil Engineers.

Torkelson, R. 1991. EPA's Environmental Financial Advisory Board: Bridging the Funding Gap. *Environmental Finance* 1(1):75–80.

———. 1991. *Incentives for Environmental Investment: Changing Behavior and Building Capital.* Washington, D.C.: Environmental Financial Advisory Board. August 9.

———. 1991. *Private Sector Participation in the Provision of Environmental Services: Barriers and Incentives.* Washington, D.C.: Environmental Financial Advisory Board. November 25.

Index

NOTE: *f.* indicates a figure; *t.* indicates a table.

ADR. *See* Alternative dispute resolution
AGC. *See* Associated General Contractors of America
Alternative dispute resolution, 120
  dispute review boards, 122
  mediation, 120–121
  summary jury trials (mini-trials), 121–122
American Arbitration Association, 116–117
American Consulting Engineers Council, 15
American Council of Independent Laboratories, 15
American Institute of Architects, 15
American Society of Civil Engineers, 15
American Water Works Association, 15
Arbitration, 116
  advantages and disadvantages, 117–118
  American Arbitration Association, 116–117
  costs of, 119
  Demand for Arbitration, 117
Associated General Contractors of America, 15, 42
  design–build standard documents, 84–85
Association of Engineering Firms Practicing in the Geosciences of Soils and Foundation Engineers, 15

Bidding, 5–6, 37. *See also* Contracts, Performance clauses, Performance incentives
  advertisement and invitation to bid, 43
  and award of contract, 44
  bid addenda, 43
  bid evaluation, 43–44
  bid form, 43
  bid opening, 43
  bid security and bonds, 43
  bidder qualification, 43
  communication during, 109
  competitive process, 42–44
  and completion date, 47–48
  consultant's responsibilities, 8
  contractor's responsibilities, 9
  documents, 13, 14*f.*, 42–43
  eliminating unacceptable bidders, 38
  instructions to bidders, 43
  invitation to bid, 43
  legal notice, 43
  owner's responsibilities, 7
  prequalification of contractors, 37–42
  regulations, 38–39
Bonds, 131–132, 132*t.*
  professional expertise needed for issuing, 134
Bonus clauses, 46
  example, 49
  maximum amounts, 49

Certificates of participation, 138–139
Chlorine contactor, 64, 65*f.*, 66*f.*
Claims
  and disputes, 112
  regarding multiple prime contracts, 75
  and warranties, 35–36
CM. *See* Construction management firms
Communication, 105
  and anger, 106–107
  basic principles, 105–108
  to clarify, 107
  clarity, 106
  conciseness, 106, 106*f.*
  concrete vs. abstract, 106, 107*f.*
  in construction phase, 109
  in design and bidding phases, 109
  to document events, etc., 107–108
  to inform, 107
  to instruct, 107
  jargon, 106
  meeting management, 108
  model, 105, 105*f.*
  and noise, 106
  in operations and maintenance phase, 109
  to persuade, 108
  in planning phase, 108–109
  purposes, 107–108
  specificity, 106
Construction. *See also* Nontraditional construction projects, Owner construction management, Traditional construction projects
  photos of progress, 57, 58*f.*, 59*f.*
  warranties, 26, 29–30
Construction Contractor's Qualification Statement for Engineered Construction, 42
Construction documents, 13–14, 14*f.*, 15*f.*, 17–18. *See also* Engineers Joint Contract Documents Committee
Construction management, xvii. *See also* Construction management firms, Design-build, Guaranteed maximum price construction management, Multiple prime contracts, Nontraditional construction projects, Owner construction management, Partnering, Traditional construction projects, Turnkey construction
  information flow, 70, 70*f.*

Construction management firms, 61, 63
  advantages of using, 63–64
  and cost savings, 63
  design consultant's responsibilities on CM project, 67
  disadvantages of using, 64
  and guaranteed maximum price construction management, 88, 89, 92, 93, 94, 95
  owner's responsibilities on CM project, 64
  payment arrangements, 63
  and quality control, 63
  responsibilities, 61, 64–67
  and risk management, 64
Construction managers. *See* Project managers
Construction phase, 6
  communication during, 109
  consultant's responsibilities, 8–9
  contractor's responsibilities, 9–10
  in design–build projects, 83
  owner's responsibilities, 7
Construction Specifications Institute, 15
Constructors
  defined, xvii
Consultants, 3–4, 3*f.*
  cost-plus fees, 11
  defined, xvii
  fees, 10–11
  lump sum fees, 10–11
  percentage fees, 11
  presentation, 10
  request for proposal, 10
  responsibilities, 7–9
  selection of, 10–11
  statement of qualifications, 10
Contract administration
  by construction management firms, 61
  defined, xv, xvii
  in multiple prime contracts, 74–75
Contractors, 3–4, 3*f.*
  and bidding regulations, 38–39
  critical characteristics, 41–42
  defined, xvii
  eliminating unacceptable bidders, 38
  prequalification, 37–42
  relationships with subcontractors, 19–20
  responsibilities, 9–10
  responsibilities in guaranteed maximum price construction management, 94, 96
  statement-of-qualification form, 42
  warranty responsibilities, 26
Contracts, 13–14. *See also* Bidding, Contractors, Engineers Joint Contract Documents Committee
  arrangements of participants, 112–113, 112*f.*, 113*f.*, 114*f.*, 115*f.*
  award of, 44
  change orders, 48
  completion date, 47–48
  cost-plus, 5
  fixed-price, 5
  performance. *See* Performance clauses, Performance incentives
  price and time changes, 20–21
  typical documents included, 15*f.*
Cost-plus contracts, 5
Cost-plus fixed-fee payment, 124
Coverage, 133–134

Delays
  compensation, 46–47
  EJCDC on, 21–22
  examples of causes, 22*f.*, 23*f.*, 24*f.*
Design–build, 11, 79–80
  compared to guaranteed maximum price construction management, 88
  competitive bid contracts, 81–82
  construction phase coordination, 83
  contracts, 81–82
  cost-plus-fee contracts, 82
  negotiated fixed-price contracts, 82
  performance parameters, 82–83
  responsibilities of participants, 80–81
  team, 80–81
Design consultants
  responsibilities in guaranteed maximum price construction management, 92, 93–94, 96
  responsibilities on projects run by construction management firms, 67
Design engineers
  defined, xvii
  warranty responsibilities, 26–27
Design phase, 4–5
  communication during, 109
  consultant's responsibilities, 7–8
  contractor's responsibilities, 9
  owner's responsibilities, 6–7
Discovery
  lack of in arbitration, 118
  in litigation, 119
Dispute resolution, 115. *See also* Disputes
  alternative modes, 120–122
  arbitration, 116–118
  costs of, 119
  dispute review boards, 122
  litigation, 118–119
  mediation, 120–121
  memorandum of agreement, 116
  mini-trials, 121–122
  negotiation, 116
  preparation for, 115–116
  summary jury trials, 121–122
Dispute review boards, 122
Disputes, 111. *See also* Dispute resolution
  caused by outside forces, 114–115
  causes of, 112–115
  over changes, 113
  and claims, 112
  over contracts, 113
  over delays, 114

inherent problems of construction, 111–112
over performance, 114
over quality, 114
regarding multiple prime contracts, 75
over site conditions, 114
Dodge Reports Bulletin, 6, 7
DRB. *See* Dispute review boards

Economic Recovery Tax Act of 1981, 133
EJCDC. *See* Engineers Joint Contract Documents Committee
Engineers
defined, xvii
Engineers Joint Contract Documents Committee, 14–16, 48
arbitration clauses in standard forms, 116
commentaries, 18–19
construction documents, 17–18
on contractor–subcontractor relationships, 19–20
on delays, 21–22
design–build standard documents, 84
engineering services documents, 16–17
modification of documents, 19
on price and time changes in contracts, 20–21
procurement documents, 18
standard documents, 14, 16
on unanticipated conditions, 22–24
Environmental problems, 38, 39*f.*
Equipment
inspection of, 30
installation of, 31–33, 31*f.*, 32*f.*, 33*f.*, 34*f.*, 35*f.*

Fast track process, 73
Final acceptance, 129
Final completion, 129
Financing, 131
attracting private-sector capital, 133
bonds, 131–132, 132*t.*
comparing public and private, 135
coverage, 133–134
general obligation bonds, 133
leasing, 138–139
and privatization, 131–133
public-sector process, 133–135
situations suited to private sector, 135–138
Fixed-price contracts, 5

General obligation bonds, 133
GMPCM. *See* Guaranteed maximum price construction management
G.O. bonds. *See* General obligation bonds
Guaranteed maximum price construction management, 87–88
appropriate projects, 89
compared to design–build, 88
compared to traditional contracting, 88–89
and construction management firms, 88, 89, 92, 93, 94, 95
construction responsibilities, 94
contractor responsibilities, 94, 96
cost management, 89–90
design and bidding responsibilities, 92–94
design consultant responsibilities, 92, 93–94, 96
objectives, 89–91
operations and maintenance responsibilities, 95–96
owner responsibilities, 91–92, 92–93, 94, 95
participant responsibilities, 91–96
planning responsibilities, 91–92
postconstruction responsibilities, 95–96
quality control, 90
scheduling, 91
variations, 88

Indianapolis Water Company
sample contract clauses, 49
Inspection of equipment, 30
Installation, 26, 28, 28*f.*, 29*f.*
of equipment, 31–33, 31*f.*, 32*f.*, 33*f.*, 34*f.*, 35*f.*

Leasing, 138–139
certificates of participation, 138–139
tax-exempt lease purchases, 138
Liquidated damages, 45–46
sample clauses, 48–49
Litigation, 118–119
costs of, 119
discovery, 119
trial, 119
Lump sum payment, 124

Manufacturers
warranty responsibilities, 26
Mediation, 120
agreement to mediate, 120–121
mediator selection, 121
process, 121
Meeting management, 108
Mini-trials. *See* Summary jury trials
MPC. *See* Multiple prime contracts
Multiple prime contracts, 69
and bid shopping, 73
cash flow management, 76
claims and disputes, 75
communication needs, 75–76
and community relations, 72
construction management approach, 71–72
contract administration, 74–75
design costs, 74
and fast track process, 73
information flow, 70, 70*f.*
and prepurchased equipment, 73–74
procedures, 76
qualifications of participants, 76

quality control, 75
and smaller contracts, 72–73
state-mandated, 69–71
total contract price, 62

National Society of Professional Engineers, 15
Nontraditional construction projects
alternative approaches, 11
warranties, 30

Operations and maintenance, 6, 33–35, 36*f.*
communication during, 109
Owner-based projects. *See* Owner construction management
Owner construction management, 11
advantages, 58–60
and approval process, 59
characteristics and activities, 53–54
construction phase, 56–57
design, 56, 59
and direct communication, 58–59
disadvantages, 60
and personnel involvement, 60
planning, 55–56
and resolution of problems, 60
staff requirements, 55, 55*t.*
staff responsibilities, 55–58
team formation, 55–56
Owners, 3–4, 3*f.*
defined, xvii
responsibilities, 6–7
responsibilities in guaranteed maximum price construction management, 91–92, 92–93, 94, 95
responsibilities on projects run by contruction management firms, 64
steps to take in turnkey construction, 84
warranty responsibilities, 25–26
Partnering, 99–100
agreement, 99, 100*f.*
case studies, 101–102
defined, 99
participants, 99–100
workshops, 102, 103*f.*
Payment, 123
combination methods, 125
cost-plus fixed-fee, 124
and front loading, 129
late, 129
lump sum, 124
methods, 124–125
and overestimating progress, 129
problems, 128–129
procedure work flow, 126, 127*f.*, 128
procedures, 123
progress payments, 47, 123, 129
and project completion, 129–130
required documentation, 127
and retainage, 128
schedules, 125–126
time and material, 125
and underestimating cash flow, 129
unit price, 124
Penalty clauses, 46
Performance clauses, 45
bonus clauses, 46, 49
delay compensation, 46–47
examples, 48–49
liquidated damages, 45–46, 48–49
penalty clauses, 46
and quality of work, 48
reduced retainage, 47
and time definition, 47
types of, 45–47
Performance incentives
and completion date, 47–48
timely progress payments, 47
Photographs
of construction progress, 57, 58*f.*, 59*f.*
Pipelines, 38*f.*, 39*f.*, 40*f.*
Pipes and piping, 64, 65*f.*, 66*f.*
Planning phase, 4
communication during, 108–109
consultant's responsibilities, 7
contractor's responsibilities, 9
owner's responsibilities, 6
Polymer feed pump, xvi*f.*
Postconstruction phase, 6
consultant's responsibilities, 9
contractor's responsibilities, 10
owner's responsibilities, 7
Privatization, 11, 131–133
attracting capital, 133
and insufficient debt capacity of government entity, 135–136
and limited transaction time, 136
and regulatory considerations, 136–137
situations suited to, 135–138
with weak government credit and strong private credit, 137–138
Professional Engineers in Private Practice, 15
Progress payments, 47, 123, 129
Project completion, 129–130
final acceptance, 129
final completion, 129
substantial completion, 129
Project engineers. *See* Project managers
Project management firms, 11
Project managers, xvii. *See also* Site representatives
office, 7, 8*f.*
Public-sector financing, 133–135
Pump stations, 62*f.*, 90*f.*

Reservoirs, 80, 80*f.*, 81*f.*, 82*f.*
Resident engineers. *See* Project managers
Resident managers. *See* Project managers
Retainage, 47, 128
reduced, 47

Safety regulations, 54, 54*f.*
Scaffolding, 54, 54*f.*

Shop drawings, 30
Site representatives, 66
Storage, 30–31, 31*f.*
Subcontractors
  defined, xvii
Substantial completion, 129
Summary jury trials, 121–122
Suppliers
  warranty responsibilities, 26
Surety, 113

Tanks, 80, 80*f.*
Tax-exempt lease purchases, 138
Tax Reform Act of 1986, 133
TELP. *See* Tax-exempt lease purchases
Time-and-material payment, 125
Traditional construction projects, 3–4
  advantages, 11–12
  bidding phase, 5–6
  compared to guaranteed maximum price construction management, 88–89
  construction phase, 6
  consultant, 3–4, 3*f.*, 7–9, 10–11
  contractor, 3–4, 3*f.*, 9–10
  design phase, 4–5
  disadvantages, 12
  owner, 3–4, 3*f.*, 6–7
  participants, 3–4, 3*f.*
  phases, 4–6
  planning phase, 4
  postconstruction phase, 6
Turnkey construction, 11, 83
  contracting for, 84
  owner's responsibilities and procedures, 84
  responsibilities of participants, 83–84
  standard documents, 84–85
  team, 83–84

Unit price payment, 124
US Army Corps of Engineers, 99
Utilities
  publicly owned vs. privately owned, xvii

Valves, 4, 4*f.*

Warranties, 25
  claims, 35–36
  construction, 26, 29–30
  elements of, 27
  equipment and materials, 26, 27–29
  and inspection, 30
  and installation, 31–33, 31*f.*, 32*f.*, 33*f.*, 34*f.*, 35*f.*
  and operations and maintenance, 33–35, 36*f.*
  participant responsibilities, 25–26
  requirements established by design engineer, 26–27
  and shop drawings, 30
  and storage, 30–31, 31*f.*
Water treatment plants, xvi*f.*, 4, 5*f.*, 27*f.*, 62*f.*, 73, 74*f.*, 89*f.*
  foundation work, 73, 73*f.*, 91, 91*f.*
Waterline, xv*f.*

AWWA Manuals

M1, *Water Rates*, Fourth Edition, 1991, #30001PA
M2, *Automation and Instrumentation*, Second Edition, 1983, #30002PA
M3, *Safety Practices for Water Utilities*, Fifth Edition, 1990, #30003PA
M4, *Water Fluoridation Principles and Practices*, Fourth Edition, 1995, #30004PA
M5, *Water Utility Management Practices*, First Edition, 1980, #30005PA
M6, *Water Meters—Selection, Installation, Testing, and Maintenance*, First Edition, 1986, #30006PA
M7, *Problem Organisms in Water: Identification and Treatment*, Second Edition, 1995, #30007PA
M9, *Concrete Pressure Pipe*, Second Edition, 1995, #30009PA
M11, *Steel Pipe—A Guide for Design and Installation*, Fourth Edition, 1989, #30011PA
M12, *Simplified Procedures for Water Examination*, Second Edition, 1977, #30012PA
M14, *Recommended Practice for Backflow Prevention and Cross-Connection Control*, Second Edition, 1990, #30014PA
M17, *Installation, Field Testing, and Maintenance of Fire Hydrants*, Third Edition, 1989, #30017PA
M19, *Emergency Planning for Water Utility Management*, Third Edition, 1994, #30019PA
M20, *Water Chlorination Principles and Practices*, First Edition, 1973, #30020PA
M21, *Groundwater*, Second Edition, 1989, #30021PA
M22, *Sizing Water Service Lines and Meters*, First Edition, 1975, #30022PA
M23, *PVC Pipe—Design and Installation*, First Edition, 1980, #30023PA
M24, *Dual Water Systems*, Second Edition, 1994, #30024PA
M25, *Flexible-Membrane Covers and Linings for Potable-Water Reservoirs*, First Edition, 1987, #30025PA
M26, *Water Rates and Related Charges*, First Edition, 1986 Second Edition 1996, #30026PA
M27, *External Corrosion—Introduction to Chemistry and Control*, First Edition, 1987, #30027PA
M28, *Cleaning and Lining Water Mains*, First Edition, 1987, #30028PA
M29, *Water Utility Capital Financing*, First Edition, 1988, #30029PA
M30, *Precoat Filtration*, Second Edition, 1995, #30030PA
M31, *Distribution System Requirements for Fire Protection*, Second Edition, 1992, #30031PA
M32, *Distribution Network Analysis for Water Utilities*, First Edition, 1989, #30032PA
M33, *Flowmeters in Water Supply*, First Edition, 1989, #30033PA
M34, *Alternative Rates*, First Edition, 1992, #30034PA
M35, *Revenue Requirements*, First Edition, 1990, #30035PA
M36, *Water Audits and Leak Detection*, First Edition, 1990, #30036PA
M37, *Operational Control of Coagulation and Filtration Processes*, First Edition, 1992, #30037PA
M38, *Electrodialysis and Electrodialysis Reversal*, First Edition, 1995, #30038PA
M41, *Ductile-Iron Pipe and Fittings*, First Edition, 1996, #30041PA
M44, *Valve Installation and Maintenance*, First Edition, 1996, #30044PA

Printed in the United States
42371LVS00002B/201-224

9 780898 678710